Integrated Pest Management for Collections

Proceedings of 2011: A Pest Odyssey, 10 Years Later

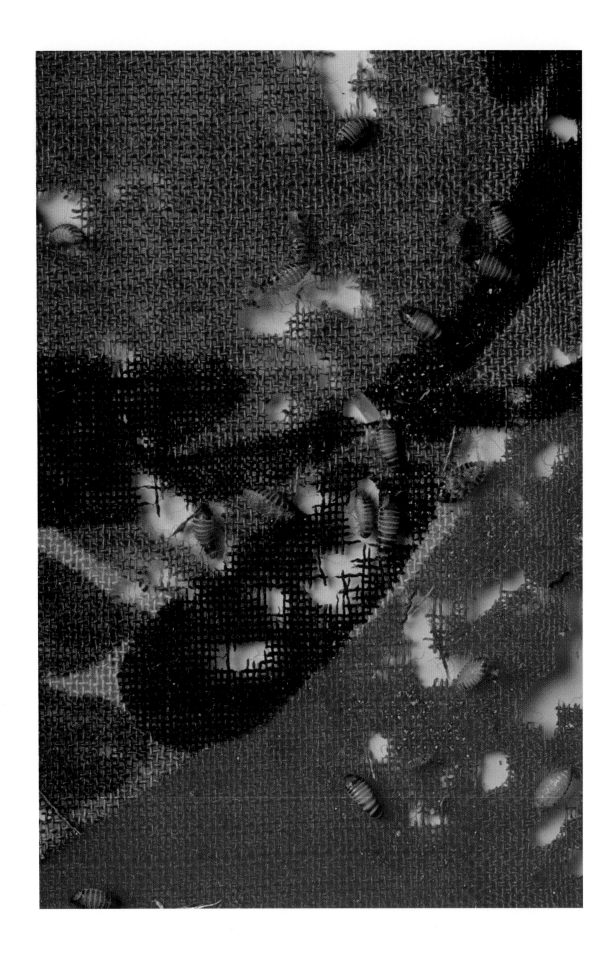

Integrated Pest Management for Collections

Proceedings of 2011:
A Pest Odyssey, 10 Years Later

EDITED BY: Peter Winsor, David Pinniger, Louise Bacon, Bob Child, Kerren Harris, Dee Lauder, Julie Phippard and Amber Xavier-Rowe

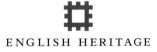

ENGLISH HERITAGE

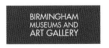

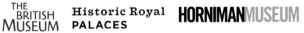

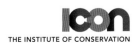

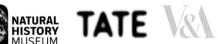

Published by English Heritage, The Engine House, Fire Fly Avenue, Swindon
SN2 2EH
www.english-heritage.org.uk
English Heritage is the Government's statutory adviser on all aspects of the historic
environment.

© English Heritage 2011

First published 2011

ISBN 978 1 84802 114 3

Product code 51706

British Library Cataloguing in Publication data
A CIP catalogue record for this book is available from the British Library.

Typeset in Bitstream Charter 9.5pt by Bookcraft Ltd, Stroud, Gloucestershire
Printed and bound in Great Britain by
Henry Ling Ltd, The Dorset Press, Dorchester

Frontispiece
Woolly-bears, the larvae of the carpet beetle
Anthrenus verbasci, *feasting on woollen fabric.*

CONTENTS

ABSTRACTS OF POSTER PRESENTATIONS

INTRODUCTION

Pest Odyssey 2011: Ten Years Later will mark an important milestone to document the changes that have taken place since the first Pest Odyssey conference held at the British Library in 2001. That conference was the first to bring together people from all over the world to discuss working with integrated pest management (IPM) in museums, libraries, galleries and historic houses. The proceedings of the 2001 conference has proven to be a benchmark publication providing a valuable tool which documented some very practical and innovative ways of dealing with pests.

There have been many changes in the last 10 years, with probably the greatest being the revolution of information technology. It is now so much easier and quicker to communicate across the world to exchange ideas, images and information. The web is also an amazing source of information on any subject from insect identification to historic use of pesticides. Even the production of these proceedings has been a totally different process from the publication process in 2001. This ability to communicate with people around the world is becoming increasingly important as pests become more widespread and the methods available to combat them become more restricted. Some of the papers in these proceedings are testament to international collaboration on a wide range of pest management topics from pest distribution to assessment of risk priorities. One of the significant themes of the papers in the 2011 conference is that they demonstrate how IPM is becoming embedded in the practices of caring for and managing our cultural heritage collections and buildings. IPM is also being taken up by smaller institutions (many run by volunteers) as seen in the Renaissance at Work (RAW) programme in England. A key reason for the acceptance of IPM is that it is now increasingly recognised that an IPM programme is more cost-effective than risking losing collections to pest infestations.

Current research is giving us a better understanding of the techniques we are already using for pest management and how this will refine our approach to treatment. There are also some novel approaches to detection and control of pests which, although at an early stage of development now, may become standard practice in the next 10 years. In view of the particular interest in DNA, the work on damage to DNA from various pest treatments is of increasing importance.

The conference in 2011 is being supported by some familiar names who presented papers in 2001, but more importantly, there are many papers and contributions from those people who either attended the 2001 conference or subsequently read the proceedings. It is hoped that Pest Odyssey 2011 will have the same influence and leave a similar legacy.

There is still much to learn and there are still new techniques to develop, as well as the results of climate change to fathom. We hope that we can look forward to another Pest Odyssey conference in 2021!

2011 Editorial group

Ten years on – from vodka beetles to risk zones

David Pinniger

DBP Entomology, Berkshire, UK, pinniger@globalnet.co.uk

ABSTRACT

The Pest Odyssey conference held at the British Library in 2001 was a milestone. It brought together people working with integrated pest management (IPM) in museums, libraries, galleries and historic houses. Ten years on from that meeting, it is time to review the progress that has been made. IPM is now accepted as the most effective way of preventing pest problems in collections. However, implementing an IPM programme in a large museum or collection can seem daunting and may result in IPM not being accepted or adopted. The concept of using risk zones was developed to evaluate and set priorities for preventing damage to collections across the whole museum. The Natural History Museum in London was the first national museum to accept the need for a museum-wide IPM strategy based on the concept of risk zones. Risk zones have also been successfully adapted for use in the IPM programmes of the Imperial War Museum, the Museum of London and the Victoria and Albert Museum. The risk zone concept has proven to be a simple but cost-effective system which can be adopted by any museum, archive or historic house to implement their IPM programmes.

Museum buildings are always likely to have a resident population of pest insects, and in the last 10 years there has been a spread in the UK of some species such as Guernsey carpet beetle, *Anthrenus sarnicus*, and brown carpet beetle or vodka beetle, *Attagenus smirnovi*. We have also seen the establishment of the dermestid beetle, *Trogoderma angustum*, in a major herbarium, and in 2011 the first infestation of Australian carpet beetle, *Anthrenocerus australis*, was documented in a UK museum. There has been a large increase in numbers of webbing clothes moth, *Tineola bisselliella*, in the UK over the last 10 years. This species has caused major problems in a number of museums and historic houses on a scale not previously experienced. The availability of cheap and effective pheromone traps for *Tineola* has enabled us to track and pinpoint the source of infestations. One unexpected conclusion is that moth infestations now seem to thrive in organic debris that has accumulated in dark and undisturbed voids. This fact has serious implications for housekeeping regimes and also the unsuitable design of many displays.

There have been changes in emphasis in control treatments and, although low temperature is still the most widely used and accepted technique for treating objects, there are a number of exciting developments. Using insect pheromones to limit populations of webbing clothes moths is now a possibility with the Exosex CLM male confusion technique. For the future, we may have to look more closely at issues such as climate change and the energy budgets of IPM prevention and control measures. What is clear is that the revolution in technology over the last 10 years has led to dramatic improvements in communication. Rapid and easy exchange of information and the availability of online training resources can only help us to improve the care of collections in our rapidly changing world.

KEYWORDS

Pests, pest management, IPM, pheromones, insect traps, clothes moths, carpet beetles

Introduction

The Pest Odyssey conference held at the British Library in 2001 was a milestone in many ways. It brought together people who were working with integrated pest management (IPM) and wanted to spread the message to others working in museums, libraries, galleries and historic houses. The *Proceedings of 2001: A Pest Odyssey* (Kingsley *et al* 2001) recorded many examples of IPM in practice and also developments and research in progress. Ten years on from that meeting it is timely to review the progress which has been made and examine any changes, predicted or unexpected.

Insects

Vodka beetles and others

As a lifetime entomologist, my first topic has to be the insects themselves. What has happened to the pests; have they increased in numbers, distribution and diversity? As I reported in 2001 (Pinniger 2001), a number of species were relatively new to the UK as pests and there were already indications that Guernsey carpet beetle, *Anthrenus sarnicus*, and brown carpet beetle or vodka beetle, *Attagenus smirnovi*, were on the increase. Since then *A. sarnicus* has spread across the UK, but infestations are still far more common in London and the south-east than elsewhere. Vodka beetle, *A. smirnovi*, has a more puzzling distribution. It has become established in many museums and buildings in London where the larvae have become entrenched in organic debris in the buildings and are exceedingly difficult to control (Pinniger 2010). The success and proliferation of vodka beetles are usually a sure sign of poor housekeeping and current cutbacks in services are bound to favour this species. It is far less travelled than *A. sarnicus*. In 2001, the only known infestation outside London was from the Fitzwilliam Museum in Cambridge. It is still there, and not only has it increased in numbers in recent years, but it has spread to the nearby Scott Polar Institute, the Sedgewick Museum and the Natural History Museum (NHM). I suspect this pest hitched a lift on a loan item from London in 1995, but has since spread in the centre of Cambridge because of the strong flying habit of the adult

beetles in warm weather. Despite the fact that it has now been illustrated in some books (Pinniger 2008) and the pest poster *A Helpful Guide to Insect Pests Found in Historic Houses and Museums* (English Heritage 2008), it has not yet been reported from anywhere else in the UK. However, I did find one dead adult vodka beetle on a windowsill in the Ashmolean Museum this year, so perhaps it has spread from Cambridge to Oxford.

Another species which was relatively new to the UK in 2001 was the dermestid beetle, *Trogoderma angustum*. Although well known to our Swedish colleagues as a pest in the Stockholm NHM, in the UK it was found only in Edinburgh and the occasional adult specimen in London. In 2003 this species was found in the herbarium at the Royal Botanic Gardens, Kew and since that time has spread around the buildings and caused some damage to dried fungal and plant collections (Pinniger and Harvey 2007). Some specimens have been found in a museum in Cambridge, but it has not yet been reported from other herbaria in the UK.

One species that I did not predict or even mention in 2001 was the Australian carpet beetle, *Anthrenocerus australis* (Fig 1). There were sporadic records in the UK (Bezant 1957) and in 1986 I found a few adult beetles in an old brush factory in Islington, London that was being considered for use as a museum store by the British Museum. In 2009, a few adult beetles were found on traps in the Victoria and Albert Museum (V&A) and also at the British Museum. In 2010, an increased number of adult beetles were trapped in the V&A and the source was eventually found to be an infested carpet in an office (Pinniger 2010). Baited dermestid larval traps were very successful in catching large numbers of live larvae (Fig 2) before the carpet was removed and the area sprayed with insecticide. An adult *Anthrenocerus* was found in the summer of 2010 on a trap in the adjacent NHM. The next year or so will be crucial to see if this pest has succeeded in spreading to other parts of the V&A and to other museums in the area. It will also be important to see if the very active larvae can cause similar damage to *Anthrenus* sp.

Moths

Small moths are generally more difficult to identify than beetles, and so identification of new or uncommon species can be difficult. The

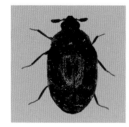

Fig 1
Australian carpet beetle,
Anthrenocerus australis,
adult.
(© DBP Entomology)

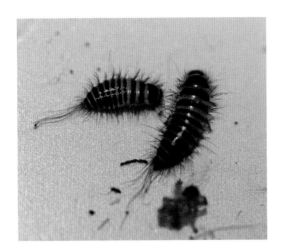

spectacular increase of webbing clothes moth, *Tineola bisselliella*, is reported later in the paper. Other tineid moth species, such as *Niditinea fuscipunctella*, have occasionally been found on traps with webbing clothes moth pheromone lures, but with no particularly apparent pattern. The widespread use of the Killgerm Demi-Diamond AF trap in heritage buildings and collections (Pinniger 2010) has increased our ability to pick up webbing clothes moths, but also seems to have enabled us to spot some new species. The tineid moth, *Monopis obviella* (Fig 3), is known to occur in bird nests, but is not regarded as a pest and has no common name. Over the last few years, staff checking moth traps have noticed an increasing number of these distinctive moths on pheromone traps. They have now been recorded from the British Museum, the Imperial War Museum (IWM),

Hampton Court Palace and Syon House. The key question is, can this moth make the transition from living on protein in bird nests to attacking wool, fur and feathers in collections? It remains to be seen whether *Monopis* can follow in the path of *Tineola* and *Tinea* and become a serious pest.

As mentioned previously, AF pheromone traps have given us a relatively inexpensive and extremely effective tool for monitoring infestations of webbing clothes moths. In addition to being essential for early warning of infestation, they have been used to pinpoint sources of infestation and carry out targeted remedial action (Blyth *et al* 2011; Baxter 2011). Consistent trap data is also vital to plot trends and detect changes in patterns of infestation. What is clear is that there has been a massive increase in problems caused by *Tineola* over the last 10 years. In 2001, clothes moths were absent or found only in very small numbers in most of the major museums in London, although there were some serious problems in some historic houses elsewhere in the UK. Populations have increased to such an extent that there are now major infestations of moths in most of the major London museums including the British Museum, the NHM, the Museum of London, the V&A and the IWM. Accurate and consistent data from the trapping programme at Hampton Court Palace provides a classic example of the spread and increase of this pest (Fig 4). From zero moths recorded in 2002, we can see a steady increase in numbers of moths caught on non-pheromone traps over

Fig 2 (left)
Australian carpet beetle, Anthrenocerus australis, *larvae.*
(© DBP Entomology)

Fig 3 (bottom left)
tineid moth, Monopis obviella, *adult.*
(© DBP Entomology)

Fig 4 (below)
Bar chart showing catch of webbing clothes moths on non-pheromone traps at Hampton Court Palace over a nine-year period.

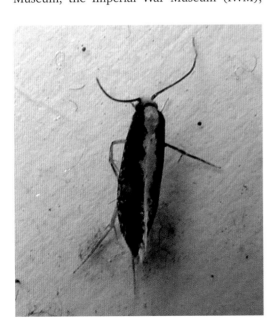

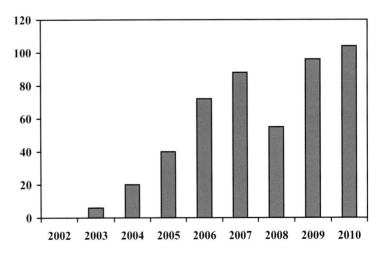

the last eight years. There has also been a very large increase in the incidence of webbing clothes moths in domestic houses (Pinniger 2009; R E Child pers comm).

Explanations for this general increase in webbing clothes moths are more difficult to understand. We can suggest possible reasons including warmer winters, more widespread use of natural fibres – including wool insulation material – and loss of potent insecticides such as DDVP/dichlorvos/Vapona. However, a more likely, although unexpected, reason for the recent success of moths in museums is the newly discovered ability of this species to live in and exploit organic debris found in dead spaces and voids in displays and buildings. In the past, it was the perceived wisdom of many authors (including me) that *Tineola* larvae could not live and breed in dirt and debris. Whether it is evolution or revolution, the moths have now proved us wrong. Well-established infestations of webbing clothes moths have been found inhabiting debris-laden voids in a number of museums and historic houses. The adult moths, which originate from populations living behind displays and under floorboards, are free to fly around and put collections at risk. Finding these populations and targeting them has been key in the success of IPM strategies (Blyth 2011; Baxter 2011 and Higgs 2011).

IPM risk zones

The concept

A number of papers presented at the Pest Odyssey conference in 2001 discussed ways of effectively implementing IPM in collections (Kingsley *et al* 2001). The concept of risk zones evolved as a result of discussions at this meeting. The first museum to take the concept through to complete implementation was the NHM in London. After extensive discussion, it was agreed that there are some key components of the system:

- recognise that most museum buildings and historic houses are always likely to have a resident population of pest insects

- evaluate the vulnerability of the collections or objects to insect attack

- assess the risk of collections being attacked by pests

- assign each area in the museum to one of four risk zones:

  A – very high (RED)

  B – high (ORANGE)

  C – low (YELLOW)

  D – very low, or none (GREEN)

- determine appropriate protocols for trap monitoring, inspection and cleaning for each zone.

The success of the project depended upon the appointment of an IPM co-ordinator with regular input from an IPM team representing the different departments. As they walked through the museum, staff from collections management teams used floor plans to identify collection risk and colour-code each area. Other points taken into consideration were: contents of cupboards in storage areas; doors, corridors and access routes; potential zone boundaries and areas with shared ownership and responsibility. It was also noted where there had been past or current pest problems. Although the priority was to map the risk zones for insect pests, it became clear that a parallel system was needed for rodent infestation. Examples of the parameters for insect risk zones A–D are given in Table 1 and rodent risk zones A–D in Table 2. It can be seen that we needed to include an additional insect risk zone 'E' for the Dermestarium. This is a special case unlikely to be encountered outside a large NHM. An example of insect risk zones marked on a floor plan is shown in Figure 5. The walking and mapping exercise was very productive, particularly when staff from different sections worked together to identify issues or problems.

After the initial exercise, it was decided to test the risk zone system in a very high-risk collections area – the Zoology Department Mammal Tower. A detailed description of this is given in Doyle *et al* (2007). The pilot study showed that the risk zone concept was both effective and practical, and therefore the museum undertook full-scale implementation. A key factor in the success of implementation has been the training of staff in IPM and pest awareness. Risk zones have had to be continually revised in the light of building changes in the museum over the last few years. Further developments include a standard trapping form and annual analysis of trapping data.

Applying the risk zone concept in other museums

Imperial War Museum

The implementation of IPM in the IWM in London was in parallel with the risk zone project at the NHM. The value of the risk zone concept was recognised at an early stage by the collections management team at the IWM. Risk zone plans were initially prepared for the IWM Lambeth Road site. They were then extended to include such diverse sites as HMS *Belfast*, the Cabinet War Rooms, IWM North and IWM Duxford. The adaptability of the concept to such diverse sites shows that the risk zone system has a potential value to all collections areas. The first priority at the IWM has been to devise an insect trapping programme to regularly collect data, and then analyse it to show changes in pest incidence and abundance. The distribution of traps is determined by the A and B risk zones at each site. This programme has resulted in effective early warning of pest problems and enabled prompt targeted action.

An outbreak of moths in the 1940s house exhibit led us to reassess the risk zones so as to include all open displays as a high-risk zone if they contained wool and other vulnerable materials. Including only accessioned objects can lead to missing a pest problem, such as this one. The issue was reinforced when an outbreak of

Table 1 Risk zone parameters for insects at the Natural History Museum, London

Zone	Insects
A (red)	Very vulnerable material (dried plants, animal skins and insects) in storage.
B (orange)	Very vulnerable material (dried plants, animal skins and insects) on display.
C (yellow)	Very vulnerable material in transit or less vulnerable material such as paper, books, osteology, molluscs etc on display or in store
D (green)	All other areas.
E (blue)	Dermestarium.

Table 2 Risk zone parameters for rodents at the Natural History Museum, London

Zone	Rodents
A (red)	All areas where food is prepared, stored, served or consumed.
B (orange)	All areas where there is public access.
C (yellow)	Offices, labs and other areas.
D (green)	Museum storage.

clothes moths was found in uniforms on figures in the open displays in the Land Warfare Hall at Duxford. We have now revised the risk zones plans and trapping programme at all the IWM sites to include all areas with non-accessioned but vulnerable objects.

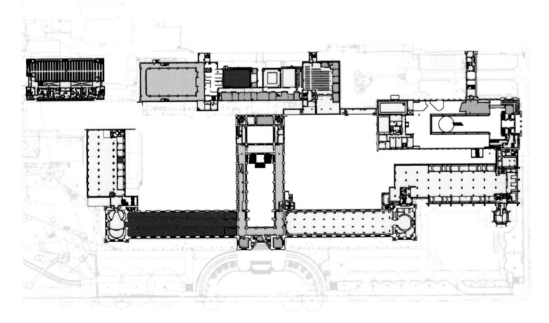

Fig 5
Risk zone plan for insects for the first floor at the Natural History Museum, London in 2007.

As with the NHM, the success of implementation at the IWM has been due to the amount of time and effort spent in training staff in IPM and pest awareness. It also relies on the effort made by staff to make sure they record the trap data accurately and on time. Continuity of staff carrying out a regular trap check has also been important to increase staff confidence and the reliability of records. The discovery of the first record of vodka beetle, *Attagenus smirnovi*, at the IWM was as a result of a staff member noticing and keeping a trap holding a beetle which looked different from the ones on the English Heritage pest poster.

Victoria and Albert Museum

Textile conservation staff at the V&A museum were one of the first groups to realise the importance of IPM in protecting collections from pest attack (Hillyer and Blyth 1992). A comprehensive trapping programme was then used in areas where textiles were stored and displayed, both on the main South Kensington site and at the Blythe House store. Trap records showed the changes in distribution of Guernsey carpet beetle, *A. sarnicus*, and vodka beetle, *A. smirnovi*, over the years (Pinniger *et al* 1998). Other departments with vulnerable collections, such as the Furniture and the Theatre Museum, also recognised the value of IPM and adopted a trapping programme. Regular pest group meetings were essential to exchange information and co-ordinate responses to any pest outbreaks. A review of collections care and functions roles at the V&A led to the creation of a preventive conservation post with responsibilities for IPM. For the first time, pest prevention could be prioritised and co-ordinated across the whole V&A museum (Smith and Blyth 2005). As with the IWM and NHM, the practical way to understand this complex site and achieve effective IPM with finite resources was to apply the risk zone concept (Fig 6).

The importance of recognising the risk to collections on open display from non-collections objects was demonstrated by an outbreak of webbing clothes moths in the Bed of Ware in the British Galleries. The source of the infestation was found to be wool textiles in the handling collection in a store in the same gallery (Blyth *et al* 2008). All vulnerable collections on open display are now assigned to the highest risk zone A.

Further development

A number of museums, including the Museum of London, have now evaluated the risk zone concept and accepted that it is a practical and cost-effective way of implementing IPM (Doyle 2011). The principle is simple, the application should be straightforward and the system fits in well with the current management concepts. Risks, priorities and cost-effectiveness are all topics that help a project to succeed. Our objective is to care for collections by preventing pest damage, and risk zones are a direct way of achieving this.

The system of risk zones can also be used in a wider context. Strang and Kigawa (2006) have devised a system for ranking the risk to collections by building type and level of protection from pest attack. They propose a series of seven categories from 'no resistance to pests' to 'comprehensive control'. A combination of this system for the overall building, together with the use of risk zones for collection areas, would give a comprehensive basis for effective IPM.

Control methods

There have been some important changes to the availability of insecticides in the past 10 years. The loss of DDVP/dichlorvos has already been mentioned, and a number of other residual insecticide active ingredients, such as chlorpyriphos, can no longer be used in the European Union. The most influential loss has been the fumigant gas methyl bromide. The banning of this gas has caused major problems to pest control in other areas, but thanks to the foresight shown by conservators and others in the museum world, it had already been largely replaced by much safer alternatives.

The situation as reviewed by Child in 2001 is now generally accepted and established worldwide. Low temperature treatments of 18°C or –30°C are the norm for the majority of collections. Anoxia is used for more fragile or vulnerable objects, but the practicalities and logistics of keeping an oxygen-free bubble for three to five weeks make it a less attractive option. There are still some problematic objects, such as carriages or grand pianos, which occasionally require treatment. First choice for these would normally be carbon dioxide fumigation, but the registration restrictions, which limit its use to one or two pest

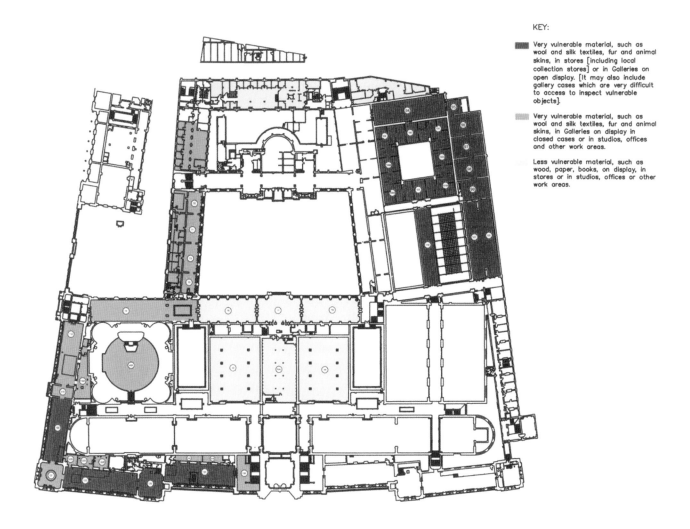

KEY:

▦ Very vulnerable material, such as wool and silk textiles, fur and animal skins, in stores [including local collection stores] or in Galleries on open display. [It may also include gallery cases which are very difficult to access to inspect vulnerable objects].

▦ Very vulnerable material, such as wool and silk textiles, fur and animal skins, in Galleries on display in closed cases or in studios, offices and other work areas.

Less vulnerable material, such as wood, paper, books, on display, in stores or in studios, offices or other work areas.

Fig 6
Risk zone plan for insects for the first floor at the Victoria and Albert Museum in 2007.

control companies in Europe, make it less attractive in the UK than in countries such as the USA and Canada where treatments can be safely carried out by trained museum staff. Issues with registration of anoxic treatments in the European Union also need clarifying.

One treatment option that I expected to become more widely used is the use of controlled heat and humidity. Tom Strang clearly stated some of the advantages of heat over freezing in 2001, but it has been little used apart from treatments carried out in Europe using the Thermo Lignum controlled heat and humidity chambers. Is this because some conservators are too conservative? I expect this topic will be hotly debated in the near future with the need to re-evaluate risks in the light of limited resources and a focus on priorities and practicalities. We urgently need more data on energy budgets because carbon footprints are becoming increasingly influential in

decision making. It would seem that preventive IPM has to be more cost-effective and energy efficient than expensive remedial treatments, but do we know if heat is greener than freezing or anoxia?

Pheromones for control

Biological control methods are frequently promoted as a replacement for insecticide treatments and provide a way forward for the future. Although there have been some undoubted successes in crop protection and horticulture, methods involving pathogens, predators and parasites are much less appropriate or practicable in museums and historic houses. A proposal to release parasitic wasps to control carpet beetle larvae in a museum received a very negative reception. One area, however, that has achieved some success is the use of pheromones, based on evidence that the presence of large amounts of female webbing clothes moth pheromone on traps had led to a reduction in moth numbers. The development by Exosect of the Exosex CLM tab system, which results in confusion of male moths and reduction in viable egg laying by females, has given us a new tool for limiting moth populations. Details of recent and current trials are given in Lauder (2011) and Higgs *et al* (2011). Understanding the behaviour of adult moths and the availability of commercially economic quantities of a potent pheromone for this species are key factors in the success of the Exosex system. Although a very exciting tool for *Tineola*, it is unlikely that this approach can be applied to other museum pest species in the near future.

Where are we going in the next 10 years?

It is dangerous to try and predict events in a rapidly changing world, but the improvements in communication over the last few years will have major implications for the future. Tools to help us identify pests and damage, and make decisions on appropriate action, are essential. The pest poster first produced by English Heritage in 1998 was updated and improved to include new species in 2008. This is now available online at www.english-heritage.org.uk together with a trap record spreadsheet, IPM guidelines and data sheets. Collections Trust published *Pest Management: A Practical Guide* in 2008 and also hosts a series of pest factsheets online at www.collectionstrust.org.uk. The release of a trial CD, *What's Eating Your Collections?*, in 2009 gave us valuable experience in preparing a tool about implementing an IPM programme. This will be the forerunner of a more sophisticated data resource that will be available online (Thompson Webb 2011). The international aspect of sharing IPM research, views and data is becoming an increasingly important part of the way we do things. The museum pests website of the Museum of Natural History in New York is a valuable source of information, and the conference at the British Museum in 2011 is helping to cement relationships with this group and colleagues in other countries.

If I have one particular personal 'Holy Grail', it is to have a reliable and accurate database of pest incidence. We have made

some progress in the UK with a pilot scheme in the West Midlands, using data contributed from sources including English Heritage, The National Trust and other museums and historic houses in that region. Such a scheme relies on three things: accurate and reliable identification of species, accurate and reliable trap data and, most importantly, the willingness of staff to collect and share data. If we can demonstrate the viability and practicality of such a scheme, we can then expand and make the programme nationwide. Then we will have a database that can be built upon and used to measure trends and changes in pest presence and incidence in the future. At last we may have data that show if there are changes attributable to climate change and other factors. The next step would be to link this with data from other European countries, the USA and elsewhere to help us in our quest for improved IPM and care of collections.

References

Baxter, P 2011 'Using Killgerm AF *Tineloa* Pheromone Lure Boards to target cleaning and intervention at the Imperial War Museum'. Poster presented at *Pest Odyssey 2011: Ten Years Later Conference Preprints, London, 26–28 October 2011*

Bezant, E T 1957 'The Australian carpet beetle in Britain'. *Entomologist's Monthly Magazine* **93**, 207

Blyth, V and Smith, S 2011 'Webbing clothes moth in the Victoria and Albert Museum's British Galleries: A successful campaign', *in Proceedings of Pest Odyssey 2011: Ten Years Later Conference Preprints, London, 26–28 October 2011*. London: English Heritage

Blyth, V, Smith, S and Chaviera, A 2008 'Actions against the clothes moth at the Victoria and Albert Museum, British Galleries'. Pest Management in Practice. *ICON Care of Collections Group Workshop.* Glasgow

Child, R E 2001 'Museums, libraries and archives. The pests: Their presence and future', in Kingsley, H *et al* (ed) *Integrated Pest Management for Collections. Proceedings of 2001: A Pest Odyssey, London, 1–3 October 2001*. London: James and James, 1–4

Doyle, A 2011 'Integrating IPM risk zones and environmental monitoring at the Museum of London', *in Pest Odyssey 2011: Ten Years*

Later Conference Preprints, London, 26–28 October 2011. London: English Heritage

Doyle, A, Pinniger, D and Ryder, S 2007 'Risk zones for IPM: From concept to implementation'. *Collection Forum* **22** (1–2): 22–31

Higgs, S and Bridal, J 2011 'Moths, Exosex™ and floor voids', *in Proceedings of Pest Odyssey 2011: Ten Years Later Conference Preprints, London, 26–28 October 2011*. London: English Heritage

Hillyer, L and Blyth, V 1992 'Carpet beetle: A pilot study in detection and control'. *The Conservator* **16**, 65–77

Kingsley, H, Pinniger, D B, Xavier-Rowe A and Winsor, P (eds) *Integrated Pest Management for Collections. Proceedings of 2001: A Pest Odyssey Conference, London, 1–3 October 2001*. London: James and James

Lauder, D 2011 The Exosex clothes moth system and English Heritage. Poster presentation at *Pest Odyssey 2011: Ten Years Later Conference Preprints, London, 26–28 October 2011*. London: English Heritage

Pinniger, D B 2001 'New pests for old: The changing status of museum pests in the UK', *in* Kingsley, H, Pinniger, D, Xavier-Rowe, A and Winsor, P (eds) *Integrated Pest Management for Collections, Proceedings 2001: A Pest Odyssey*. London: James and James, 9–13

Pinniger, D B 2008 *Pest Management: A Practical Guide*. London: Collections Trust

Pinniger, D B 2009 'Clothes moths – numbers are surging'. *Pest Magazine*. July/August 2009

Pinniger, D B 2010a 'Watch out! Vodka beetles about'. *Pest Magazine* September/October 2010

Pinniger, D B 2010b 'Saving our heritage: Pest management in museums and historic houses'. *Outlooks in Pest Management*, October 2010

Pinniger, D B, Blyth, V and Kingsley, H 1998 'Insect trapping: The key to pest management', *in Proceedings 3rd Nordic Symposium on Insect Pest Control in Museums, Stockholm, 24–25 September 1998*. Stockholm: Naturhistoriskariksmuseet, 96–107

Pinniger, D B and Harvey, Y 2007 'The Stockholm beetle *Trogoderma angustum*: A new risk to herbarium collections'. *NatSCA News* Issue 12

Smith, S and Blyth, V 2005 'Prevention is better than cure'. *Victoria and Albert Museum Conservation Journal* **50**

Strang, T J K 2001 'Principles of heat disinfestation', *in* Kingsley, H *et al* (ed) *Integrated Pest Management for Collections, Proceedings of 2001: A Pest Odyssey Conference, London, 1–3 October 2001*. London: James and James, 114–29

Strang, T and Kigawa, R 2006 'Levels of IPM control: matching conditions to performance and effort'. *Collection Forum* **21**(1–2), 96–116

Thompson Webb, J 2011 'But how do I know I've got pests?', *in Proceedings of Pest Odyssey 2011: Ten Years Later Conference Preprints, London, 26–28 October 2011*. London: English Heritage

2

Ten years of integrated pest management at English Heritage

Amber Xavier-Rowe* and Dee Lauder†

English Heritage, London, UK

*amber.xavier-rowe@english-heritage.org.uk   †dee.lauder@english-heritage.org.uk

ABSTRACT

Integrated pest management at English Heritage covers 65 sites and has been established over the past 10 years. Managed centrally and delivered by site-based staff, the programme has been instrumental in preventing major insect pest infestations. Catch data recorded since 1997 indicates that webbing clothes moth activity is increasing. The main sources of insect pests and preventive and treatment approaches are outlined.

KEYWORDS

Integrated pest management, IPM, insect pests, historic collections, insect monitoring, insect pest control

Introduction

Integrated pest management (IPM) at English Heritage (EH) has been instrumental in preventing damage to significant collections displayed and stored at 65 sites over the past 10 years. This is a remarkable achievement as these sites display and store vulnerable materials including wool, leather, natural history specimens, paper and wood. This paper describes how and why IPM at EH has been so successful.

Background

English Heritage is the United Kingdom government's statutory advisor on the historic environment for England. One of its key roles is the conservation and presentation of over 400 properties. There are 115 sites that display or store collections of which 65 sites house vulnerable collections including wool-based furnishings, natural history specimens, furniture, books and paper artefacts. Those collections on open display in historic buildings are the most at risk from insect pest attack.

IPM commenced at EH in 1997 starting with a sticky-trap monitoring programme at Audley End House, a 42-roomed Jacobean property, which displays and stores 22,478 objects. The developing EH IPM strategy at the time was outlined in a paper published by Xavier-Rowe and Pinniger (2001) in *Pest Odyssey 2001*. Since 2003, the IPM programme has been centralised under the management of one person, our Collections Pest Control Manager, with great success.

In the EH State of Collections Report (Xavier-Rowe and Fry 2010), the risk posed by insect pests was deemed to be low. The report was based on evidence provided from a collections condition audit and site-based risk assessment completed for 115 sites. This result confirms the effectiveness of IPM at EH as the overall risk of insect pest damage is increasing for historic house and museum collections. It is the opinion of the authors that insect pests should be considered as one of the highest potential risks for historic collections as the density of vulnerable materials on display or in store provides an ideal environment for insect pests to thrive.

The IPM system at EH

The key elements that work together to produce a sustainable and effective IPM programme at EH are described below:

Insect pest trapping and interpretation

The foundation for success at EH is a systematic monitoring system delivered by a range of people who have been coached and supported by the Collections Pest Control Manager. The monitoring system, based on sticky museum traps and pheromone lure traps, has been designed so that site staff, conservators, collections care assistants and curators can monitor the traps. Keeping the number of traps to a realistic number and checking them two to four times a year has proved to be achievable. Results are logged onto an Excel spreadsheet and house plans using a standardised key chart. These were created to enable staff to electronically send in the results by email every quarter instead of posting paper returns (Lauder 2009).

However, an element of quality control is required with 27 site-based staff completing the returns. All quarterly or bi-annual returns are checked by the Collections Pest Control Manager to remove errors and quickly spot any unusual insects or potential insect pest problems. High catch numbers are investigated either over the telephone or through a site visit. Annual insect trapping and monitoring reports are prepared for each property which highlights trends in terms of insect pest numbers and actions needed to reduce the likelihood of an infestation. The annual site report is circulated widely to both inform and raise awareness of insect pests and the ongoing actions being taken to control them.

Annual results have been gathered and recorded in this manner since 1997, providing useful trend data which has directly informed collections care practices. At Audley End, for example, the data relating to the varied carpet beetle, *Anthrenus verbasci*, webbing clothes moth, *Tineola bisselliella*, and case-bearing clothes moth, *Tinea pellionella*, flagged up issues relating to housekeeping and chimney cleaning (Fig 1). *Anthrenus verbasci* numbers decreased over seven years until 2005 when numbers suddenly increased. Upon investigation, it turned out that housekeeping standards had dropped due to staff changes. Whilst the impression was given that all was well, the deep cleaning of vulnerable rooms and collections was not being targeted effectively. The monitoring results provoked a change to the housekeeping schedule and recognition by the conservator and collections care assistants that certain areas and collections in the house needed to be deep cleaned more frequently during the summer months. The new schedule was implemented during 2006 and the catch numbers started to decrease. However, in the last two years they have increased again, which is related to a

Fig 1
Audley End House insect pest catch results 1997–2010.

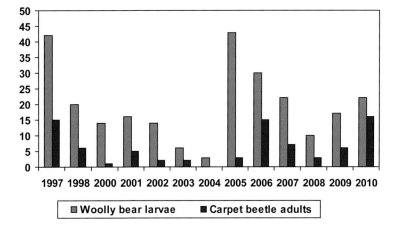

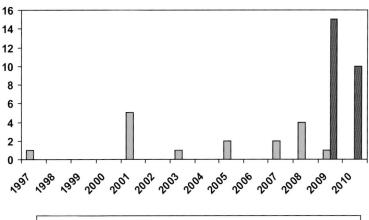

number of chimneys that require sweeping. The arrival of *Tinea pellionella* in the last two years in high numbers is also linked to debris in these chimneys. A programme of chimney cleaning has therefore been implemented.

In order to keep the monitoring programme sustainable, properties have been divided into four categories. This has ensured that effort is focused on the important and vulnerable collections. Category A and B sites (33) are monitored four times a year. Category A sites hold the most important objects whilst B sites may have less important collections that are still vulnerable to attack. Category C sites (7) are monitored twice a year, during the spring and summer months, whilst D sites (23) are annually deep cleaned and visually checked. Category D sites do not have an annual site report written up as there are no monitoring records. Most of these sites are 'buildings related' where, for example, there has been a history of wood borers in the structure or just a few vulnerable items on display such as pews and traceries in churches.

Annual site reports are written up, based upon the quarterly trapping information over the past year, and are either emailed to the individual sites and staff concerned or compiled together into a report (Lauder and Pinniger 2010). This is circulated to all the managers involved, including senior management, with the purpose of raising awareness of IPM as a long-term collection care activity. The annual site reports have been produced for the last eight years.

Centralised management

The sustainability and effectiveness of the EH IPM programme are due to the centralisation of management under one person supported by senior management. In many organisations, pest management duties are usually undertaken as an add-on to a job description. Until a dedicated post was created in EH in 2003, progress had been inconsistent and difficult to sustain. At EH the conservators and collections care assistants mainly assist with IPM, but they do not have the time to focus on monitoring, reporting and dealing with potential problems before they turn into an active infestation.

The other main advantage of having a dedicated post is that this person can keep up-to-date with key developments in monitoring and control as well as health and safety

regulations and other legislation, for example, treatments and protected species. Whilst the focus of the Collections Pest Control Manager is on insect pests, vertebrate bodies and the baits left by contractors are becoming an increasing problem to collections as they provide a food source for the insect pests.

English Heritage IPM strategy

The EH IPM Strategy was written in 2006 and last updated in 2011 (Lauder and Pinniger 2011). It is used widely by staff involved with monitoring as well as being used by the senior management as the formal set of standards for implementing IPM at our sites.

Training

At the heart of the influencing, coaching and training programme is the EH poster recently updated to include new pest species (Pinniger *et al* 2009). This simple publication has been very effective both at raising the awareness of IPM and as an insect pest identification tool.

The training programme consists of four courses. The IPM training course taught over two days concentrates on insect pest identification and gives an understanding of how they become established in historic houses and collections. An important learning outcome is to correctly identify insect pests and the damage they cause. The EH monitoring and recording system is then introduced through practical sessions. This can then be set up and established with participants over the following year through one-to-one coaching at their sites by the Collections Pest Control Manager. Since 1998 we have trained 119 members of staff.

The IPM master-class is a follow-up day course designed to provide EH house staff with updated information which advances the knowledge they have all previously gained by attending the IPM training course. It introduces new pest species and also any updates to our IPM procedures. Other topics covered include bats and legislation, and other insect pest trapping techniques currently available. The presenters provide instruction, practical sessions and advice. Since 2001 we have trained 39 members of staff and one person from the National Trust for Scotland.

The pests master-class, co-presented with vertebrate consultant Ed Allan, is for EH conservators, IPM-trained staff, building maintenance managers and property curators. Updates are given on current insect pest species and issues and their implications for the collections and buildings. Other topics covered include vertebrate issues, protected species updates and also new low-hazard/non-chemical treatments and prevention methods. All current legislation and health and safety issues are also covered. We also advise on pest control companies or consultants who are experienced in working in the historic house context. Since 2008, we have trained 45 members of staff and two external members of staff from Historic Royal Palaces.

An Insecticide Treatment course co-presented with Bob Child is also run. Conservators, collections care assistants and curators are trained in the safe use of desiccant dusts and Constrain insecticide application using pump sprays and 'fogging' equipment. The training also covers all current health and safety and legal requirements. From 2005 to the present day we have trained 22 members of staff and 12 members of staff from other heritage organisations.

National trends

On reviewing the national data over the past 14 years we are starting to get a picture of which insect pests are on the increase. Looking at results for webbing clothes moth, *Tineola bisselliella,* numbers have increased sharply since 2008 (Fig 2). Whilst the introduction of more effective moth lures in 2008 is responsible, in part, for the increased catch, they cannot be totally accountable for such a dramatic rise. These results suggest that of all the pest species, clothes moth currently present the greatest risk to EH collections.

This type of long-term data analysis can both provide a warning to the risk level and help with securing and targeting resources for research into control methods. To this end EH staff have been working with David Pinniger and Jane Thompson Webb at the Birmingham Museum and Art Gallery to provide data for an online national IPM database that could be used to highlight risk levels by region and town. The project is currently being piloted on the What's Eating Your Collections website. When we have a good baseline, with data from a wide range of reliable sources, we can use this information to show changes in distribution and frequency of insect populations and how they are affected by climate and other factors.

Sources of insect pests in English Heritage sites

Through maintaining an IPM database on which all information relating to IPM at each site is logged, we can confirm the main sources of insect pests.

Poor housekeeping

Poor housekeeping is by far the biggest contributor to increases in pest activity. The build up of dead insects, including flies, ladybirds, dirt, dust and litter, has been responsible for increased pest activity.

Chimneys

Chimneys, which are nearly always present in EH sites, are the principal source of significant rises in moth species. They have been largely missed from cyclical maintenance schedules as they are no longer used. This situation has changed at EH through the IPM programme, and chimney cleaning and capping is now recognised as a core maintenance activity.

Forgotten rooms

Rooms not open to the public are often left off cleaning schedules. These spaces often become the final resting place for dead insects, and even birds, which have then attracted insect pests.

Fig 2
English Heritage properties webbing clothes moths catch 1999–2010.

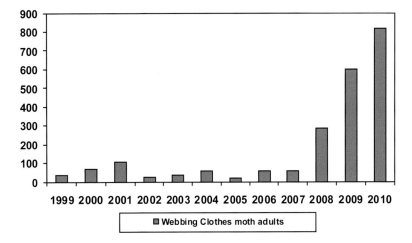

Fig 2

Lack of building maintenance

The lack of building maintenance related to downpipes, guttering, roof spaces, bird proofing, window- and door-proofing, and roof repairs have all been responsible for damp ingress resulting in death watch, *Xestobium rufovillosum*, and furniture beetle, *Anobium punctatum*, activity in the fabric of the building.

Vertebrate pests

Birds, rodents, bats and other protected species, squirrels, rabbits and moles have also been responsible for pest activity through nesting materials, droppings and dead bodies.

Prevention and control

Producing an annual report for each IPM site provides the key information for prioritising actions over the coming year and is fundamental to preventing damage.

There are about 600 chimneys in the 23 Category A and B sites that require cyclical cleaning. Chimneys that are linked to rises in insect pests are prioritised for cleaning using a budget that has been ring-fenced for collections maintenance. Requests for chimney sweeping are logged on the building maintenance database system to ensure that jobs appear on cyclical schedules using agreed specifications. Establishing a close link with Estates teams through engaging with their system should mean that this relatively simple and cheap task which can have such a major impact on collections is dealt with in a methodical and timely manner. We also alert maintenance teams of a range of building maintenance issues noticed through insect monitoring.

Housekeeping schedules are regularly reviewed and revised in response to annual results and targeted deep cleans are undertaken when required.

Birds and rodents are becoming an increasing problem for collections often due to the increased consumption of food and frequency of functions at many sites. We are therefore aiming to influence EH practices relating to vertebrate control through a standard specification for the appointment of contractors and advocating a central cyclical contract carefully monitored to ensure effective control and value for money.

Control treatments

The insecticide Constrain, a pyrethroid micro-emulsion is used for the local treatment of textiles (carpets, curtains, upholstery), plant fibres and wood.

Fogging using Constrain and the IP Mini Fogger is used to treat rooms both as a preventive measure and for control of moth outbreaks. This control measure is mainly used for large recreated interiors where wool has been used.

Temperature treatments, freezing and heating, are the preferred methods. For the treatment of multiple objects we prefer heating using the Thermo Lignum (UK) Ltd mobile treatment chamber due to the short treatment time and proven efficacy, particularly for wood borers (Strang 2001). Following the successful trial of Exosect Ltd's Exosex CLM and CL moth confusion pheromone lures since July 2007, we can now consider deploying it at other sites to control webbing clothes moth numbers to acceptable levels. This is a non-chemical 'pest confusion' treatment designed specifically to reduce the number of highly destructive larvae of the webbing clothes moth. It uses a synthetic female pheromone to attract male clothes moths into a dispenser where the 'Entostat' powder combined with the pheromone is situated. Males are lured into the dispensers and upon entering the powder coats their bodies. The senses of the coated moths are overwhelmed and they cannot detect females as a result. As they leave the dispenser, they then attract other male clothes moths and so spread the confusion effect even further. Female moths do not get mated and lay very few fertile eggs and as a consequence there are far fewer larvae.

The introduction of wool-based materials as part of new presentation schemes is carefully managed. Where possible wool is avoided, however, this can be challenging when authenticity, texture and drape of textiles are essential to the successful historic interior scheme. Where no acceptable material can be found to replace wool, the method of installation is controlled to ensure easy access for removal and cleaning. In some cases we have also implemented an annual fogging with Constrain insecticide to prevent a moth outbreak.

A significant proportion of EH collections (87%) are in store (Xavier-Rowe and Fry 2010). We are in the process of developing new storage facilities on our estate. This is an excellent opportunity to dispose of accumulated materials, check vulnerable collections as they are packed and to design the new stores so that relative humidity can be kept below 60% for most of the time. Quarantine areas and procedures for receiving goods and collections relating to insect pests are also being updated.

Raising the public profile of IPM

The insect pest story can be very successful in attracting public interest through the media. When we have given a press release relating to IPM the response has been strong. The most recent example is the in-depth interview with Dee Lauder by BBC Radio 4 as part of a programme called 'What's Eating The Museum?' about pest control in museums and historic collections in 2011 (Fig 3).

Conclusion

With climate change and the cuts to funding in the UK, the risk of major damage to the nation's heritage from insect pests is increasing. IPM successfully mitigates this risk, which has been the experience at EH over the past 10 years. It is an efficient, manageable and effective strategy at EH because one staff member is responsible full-time for the programme.

Acknowledgements

We are indebted to David Pinniger who has inspired, trained and coached the authors and staff at EH for the past 10 years.

References

Lauder, D 2009 English Heritage insect ID monitoring sheet, web version. www.english-heritage.org.uk/professional/research/heritage-science/collections-conservation/collections-care-guidance/ (accessed 27 September 2011)

Lauder, D and Pinniger, DB 2010 'Pest Management in English Heritage Properties 2010: Summary'. Unpublished internal report, English Heritage: London

Lauder, D and Pinniger, DB 2011 'English Heritage Integrated Pest Management Strategy'. Unpublished internal paper. London: English Heritage

Pinniger, DB, Xavier-Rowe, A and Lauder, D 2009 Insect Pests Found in Historic Houses and Museums [Poster]. London: English Heritage www.english-heritage.org.uk/content/publications/docs/insect-pests-historic-houses-poster.pdf (accessed 18 August 2011)

Strang, TJK 2001 'Principle of heat disinfestation', in Kingsley, H, Pinniger, D, Xavier-Rowe, A and Winsor, P (eds) *Integrated Pest Management for Collections, Proceedings 2001: A Pest Odyssey*, London: James and James, 114–29

Xavier-Rowe, A and Pinniger, D 2001 'No uninvited guests: Successful pest management in historic houses', in Kingsley, H, Pinniger, D, Xavier-Rowe, A and Winsor, P (eds) *Integrated Pest Management for Collections, Proceedings 2001: A Pest Odyssey*. London: James and James, 37–43

Xavier-Rowe, A and Fry, C 2010 'State of English Heritage Collections Report'. Unpublished internal report. London: English Heritage

'What's Eating Your Collections?' website, www.whatseatingyourcollection.com, online database of pest information and recording. Project based at Birmingham Museum and Art Gallery funded by Renaissance programme of Museums, Libraries and Archives Council. (accessed 27 September 2011)

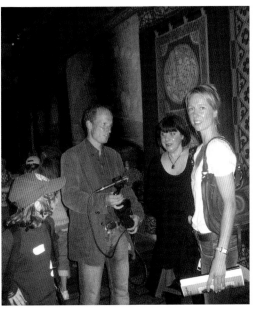

Fig 3
Dee Lauder and Ann Katrin Koester from English Heritage being interviewed for a BBC Radio 4 programme 'What's Eating The Museum?' about pest control in museums and historic collections.
(Photo © English Heritage)

3

Efficacy, effects, economics: the problem of distributing pest control advice to cover contingency

Tom Strang* and Rika Kigawa†

*Canadian Conservation Institute, Ottawa, Canada, Tom.Strang@PCH.GC.CA
†National Research Institute for Cultural Properties, Tokyo, Japan, rkigawa@tobunken.go.jp

ABSTRACT

Our organisations assist a large and diverse group of cultural institutions with conservation advice and services. To help them begin integrated pest management programmes we developed a set of guidelines for combating pests in cultural property. Firstly, pests are introduced at a basic level to convey their needs and the hazards they present to the materials found in collections and structures. Secondly, a suite of control activities are structured by their desired outcome towards pest reduction that integrates with a larger scheme for preventive conservation of collections. Thirdly, activities and features to combat pests were distributed across levels chosen to represent easily recognisable situations. The document was designed as a concise starting point for implementing pest management of cultural property, midway along the spectrum from detailed scientific understanding of pest problems to administrative oversight of cultural property.

KEYWORDS

Integrated pest management, IPM, advice, planning, training

Introduction

A canting totem pole is slowly rotting on a storm-swept ocean coastal site, and when it falls it will be allowed to rest. A weathered carriage sits under a shed roof with a persistence of moths eating its remnant wool upholstery. Ridden with holes, among piles of frass, carpenter's tools lay in abandoned illustration of your great-grandparents' DIY lives on in a struggling rural museum. Intermittently damp basement storerooms in the old city hall quietly contaminate art and objects with mould. An in-coming exhibition threatens in hurly-burly to 'Trojan-horse' well controlled exhibition halls with unknown harms. Hundreds of tatty cardboard boxes packed solid with paper and diskettes will arrive for painstaking processing into funds of knowledge in a national archive.

What to do? Who's to do it?

Pest hazards have two ends to manage. We cope with pest crises, and want to prevent them. While understanding pest biology and behaviour, and having efficacious measures required at both ends, the distributions of activities differ.

In pest crisis, actions are justified against mitigation of ongoing harm, whereas sustaining prevention effort needs measures to justify continued activity against evident lack of harm.

To be good, general preventive advice needs internal logic with the support of scientific tests. The minimum expectation is 'good advice subsumes expert opinion'. But good advice is not uniformly applicable in all contexts, not with strongly competing factors.

Cultural properties will often be cared for by people experienced in local factors, without specific training or experience in pest matters. Among this group, where and how to begin pest management activities are common questions. Even with experience, novel situations need thinking through in order to form a reasoned approach.

The pest expert provides understanding of pest needs, capabilities, vulnerabilities, and can suggest counter-actions, corollary experience, information on side effects and so forth. The collection expert provides values not to be traded for efficacy, such as effects to be avoided, restrictions on time, vetoing dislocating actions required by remedies, deciding an overall logistic and cost envelope, and so forth. Structuring both parties' knowledge is what integrates pest management (IPM).

The last twenty years have seen a shift in pest control practices, primarily from prophylactic fumigation with reactive chemical gas and pesticide use on objects, to methods more benign to the objects, with IPM approaches to whole collection stores in order to lower incidence, and emphasis on monitoring to guide response. However, collections are to be found within different cultural contexts than simply enclosed in museum cases and rooms, a scope which makes pest control recommendations an interesting challenge.

The bulk of the world's cultural resources are not in the few top museums. They are spread throughout civic archives, education facilities, regional museums, historic villages, churches, temples, shrines, private homes, and outdoor sites. IPM advice has to cover the needs of a large number of institutions across a spectrum of resource and individual capability.

It is estimated that Canada has 2,400 museums (Canadian Heritage Information Network) and nearly 4,000 museums, archives and similar institutions (Canadian Museums Association) depending on the manner counted. The Canadian Museums Association directory (1997) yields the largest breakdown by their core purpose (Fig 1).

From a survey of 1,476 museums (Statistics Canada 2002), of museums with less than

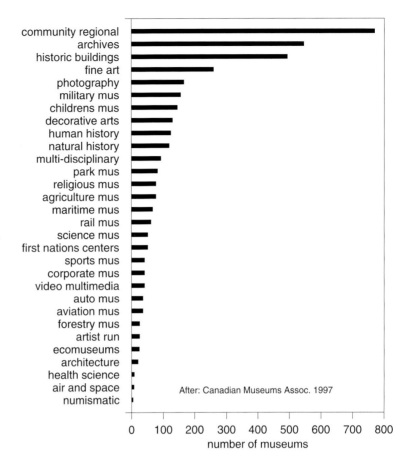

After: Canadian Museums Assoc. 1997

number of museums

Fig 1
Canadian museums by function.

$40,000 annual budget, one in ten have permanent year-round staff, and volunteer ratios of about 5:1 to paid staff (Fig 2). Over 40,000 people work in these institutions, and nearly three-quarters are volunteers.

Japan documents 5,614 museums, with 3,200 history, 1,087 art and 474 science museums. Of these, 865 are registered museums, 331 museum-equivalent facilities and 4,418 museum-like facilities. Only the first two categories, as part of their conditions, require trained curatorial staff (MEXT 2008).

The Japanese government has a designation system to nationally fund conservation and restoration of cultural properties, including buildings and fine art objects. These are designated as Important Cultural Properties (ICP) and, within this category; exemplary buildings and items have a higher designation known as National Treasures. For example, 4,404 buildings, most of which are traditional buildings of temples and shrines, are designated as ICP, and among them 264 buildings are National Treasures. Within the 10,387 fine art works under ICP auspices, 866 are

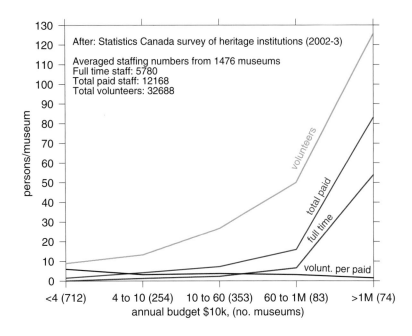

Fig 2
Staffing in Canadian museums.

checklist of things one can do is not particularly useful without some idea of how to select only the germane elements (avoid pointless dogmatism). We also want to avoid outcomes that are ineffective (wasteful effort) or greatly reduce primary functions for the sake of pest control (draconian). Detection programmes and review of the programme temper our choices over time. IPM programmes should not fail, but evolve. To similar effect, the subdivision of large institutions into zones which reflect local collection sensitivities and core activities provides flexibility so IPM programmes can be adopted and sustained (Doyle *et al* 2009).

To benefit the intended audience, the design of the CCI Technical Bulletin 29 publication (Strang and Kigawa 2009) tried to capture much of the expert guidance available, and display it in a manner relevant to the widest range of readers so that they can find a solution to their problem.

National Treasures (Agency for Cultural Affairs 2011).

Approximately 60 per cent of ICP-designated fine arts are owned by temples and shrines and these are often kept in the traditional buildings of temples and shrines which generally fall into our level three environment.

A new system of registering historical properties outside the ICP designation system has also been started to help conserve them, and more than 8,000 buildings are now listed. In addition, Japan has 88 Preservation Districts for groups of traditional buildings, which include inside their area approximately 16,500 traditional buildings and structures (Agency for Cultural Affairs 2011). Pest advice is useful to this wider population of structures which often contain locally significant items.

Requirement drives publication design

The authors' public advisory roles entail responsibility to assist cultural institutions in making decisions for protecting their holdings from pests. Structuring integrated pest management advice into comprehensible and achievable work becomes part of every conversation beyond the specific technical discussion on how to kill the pesky problem off.

Flexibility in IPM design allows people to create IPM responses commensurate with their situations. However, proceeding from a flat

Education on pests and their needs

The largest step we can make towards reducing the threat of pests to cultural property is to widely educate those who are responsible for care through fostering understanding of the problems in preventing pests, and how to break these down into manageable elements.

Introduce the essential-to-know pests and their requirements

Not all life forms pose a threat to collections or historic sites. In fact, only a small fraction of the total known of the world's species come in direct conflict with the human desire to preserve cultural property. It can be a relief to know that not every insect seen is cause for great concern.

We were also aware of regional issues around identification, and were not looking to compete with all of the identification resources already available. The beneficial growth of web-based identification sites will likely handle regional needs and languages. Japan has its own illustrated cultural pest reference (IAI NRICPT 2001), as do other countries such as the UK (Pinniger 2001, Pinniger 2008) and Sweden (Akerlund 1991).

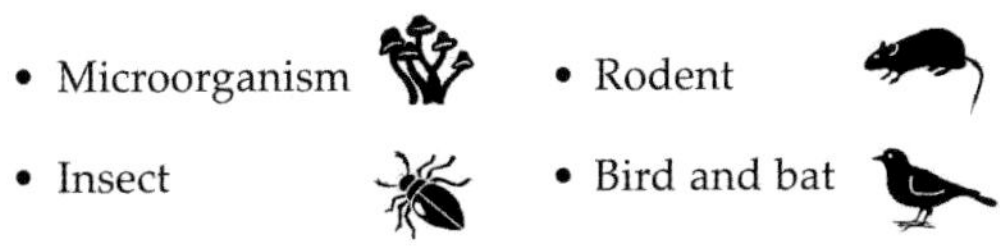

- Microorganism
- Insect
- Rodent
- Bird and bat

We chose to focus on a minimum set for Canada. This parsimony is not particularly debilitating as many of the species are widely distributed or have analogous species elsewhere. Rather, the chosen set quickly orients a novice reader to the key forms of collection pests we most commonly see.

For a quick visual index throughout the unavoidable tables of detail, we chose four icons (Fig 3) to represent microbes, insects, rodents and birds/bats, which cover most of our clients' concerns bar the occasional snake, raccoon or bear.

Emphasise the probability of harm befalling certain materials

We see the pests that time forgot. In the industrial societies of today and with our high use of synthetics, collections house a greater variety material for supporting pests than is encountered in the home. We no longer all live next door to a barn, grain mill or weaver.

Discussing the sensitivity of collections to pests is essential to help novices understand what is particularly vulnerable. Pests are consumers of materials not objects. While there is some adventitious chewing, most harm is done when specific materials are attacked as food by a pest that has the capability to digest it. Curatorial knowledge can be used to map the distribution of IPM efforts.

Education on actions that counter pests

Actions speak louder than words. Control activities were organised within categories as described in the Canadian Conservation Institute's (CCI) systematic preservation chart. This approach was refined through many pest workshops and publications for specific audiences (Mickalski *et al* 1992, Strang 1998, 1999, Kigawa *et al* 2003).

At its heart, this simply presents IPM information in the following manner:

- for what you cannot avoid, block
- for what you can detect, respond.

Illustrate with real examples

Brief case studies were made from notes of CCI consultations with organisations who had enquired about pest advice. These were picked as key examples across the spectrum of institutions, from small to large, within urban and rural environments. They are illustrative of common patterns of pest attack: rodents inhabiting structures and damaging objects, mould on objects caused by failure of the structure, harmful insects on objects coming from an uncontrolled source, and insects burrowing and damaging large wood objects and buildings. These examples give the reader some reference experience similar to what they might acquire in practice.

Application of knowledge in IPM design

With a basic understanding of pests, their collection's vulnerability, and activities that can be undertaken, it is reasonable to expect people to want to know how to go about solving their pest problems. The primary application guide kick-starts an integrated pest management plan. While the case approach provides a storyline that is truthful to a subset of experience and can instructively document the decision process used to get out of the situation, this still leaves the problem of organising 'expert knowledge' into a form that is accessible to the reader in a way that captures the sorting and emphasis that a practised decision-maker would apply in the course of dealing with any specific case.

This was the core problem looked at by the authors' 'levels of IPM' approach (Strang and Kigawa 2006). We have worked this approach into both the Technical Bulletin 29 and a Japanese language poster.

Create families of possible actions and rank sort their likely effectiveness

Activities contributory to IPM were classified into nominal categories such as containment, cleanliness, attraction and so forth, and sorted by their likely contribution toward their primary intended goal from minimum effect to maximum effect. We called these 'scales,' akin to the way ordinal clusters of observables have been used to estimate phenomena in the

Fig 3 (left)
Pest icons.

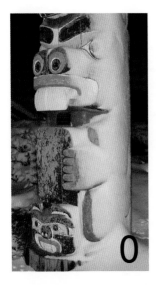
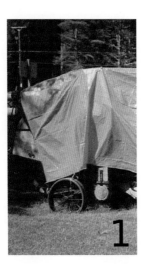
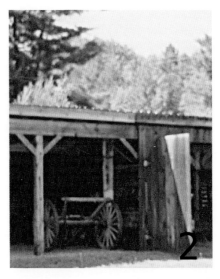
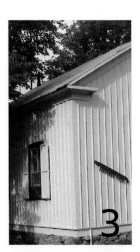
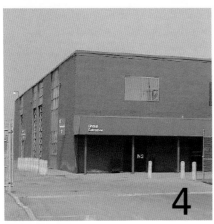
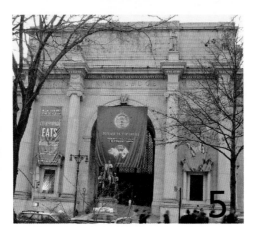
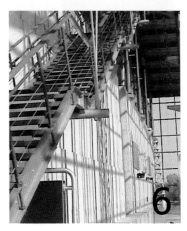

Fig 4
Visual scale of levels.

absence of a direct measure (Shindo scale to estimate earthquake strength, Beaufort scale to estimate wind strength).

While we had numerical data to inform some of our rankings, we avoided the pitfalls of numerical semi-quantitative methods of ranking and were not prepared to embark on a full quantitative analysis at the time. The AS/NZS risk analysis guidelines (2004a,b) recommend that a qualitative approach be taken as a first exercise. In making the scales we reviewed available illustrative accounts that informed this qualitative ranking exercise (Strang and Kigawa 2006). We also felt that those of our readers in a position to do the most good would not be best served by quantitative arguments.

Pick the key protective mechanism as index

Biological harm stems from organisms which were originally in open natural environments, and in the absence of human activity would likely be adventitious consumers of dispersed resources instead of singular depleters of stockpiles. 'Cultural stockpiles' range from monolithic (eg paper archive) to fully mixed (eg history collection) in their distribution of vulnerable materials, and these common operational categories all have pest-related threats of some kind.

We decided that primary containment would be the key entry point for the reader. It is what you see when you arrive for work. We called these 'levels' (Fig 4) and tried to evoke clear steps between them, although not necessarily equidistant. While 'levels' implies the concept of progress upward or downward as far as the potential for preservation against pests, we recognised up front that one may infrequently move a level of containment to gain a specific benefit (for reasons such as high cost or it being detrimental to overall objectives) so an IPM plan has to be tailored to the situation at hand – 'Plan B'. With this in mind we populated a matrix across levels retaining the rank ordering of the scales (Strang and Kigawa 2006).

Estimates of effectiveness

This was the most daunting part of the exercise and borders on quantitative risk assessment. Depending on the degree of gut-level credence people ascribe to the scales, such an estimate may not be necessary to sway arguments for gaining improvement. However, we were able to determine some measures around timescales for loss for steps where we found data on rates involving pest exposures and material survival (Strang and Kigawa 2006). We see this as a clear area for future work in IPM and risk management as it incorporates measures of efficacy, effects and economics into the decision processes.

Illustrate readily available treatments

We maintain that a large amount of good can be done with rudimentary methods (effective sanitation, timely maintenance and so forth).

How to kill insects in objects is a common request. The applications section illustrates thermal pest treatment guides as these methods handle a modest number of objects, are quickly enacted and efficacious. The instructions were conceived as one to two sheet graphic guides (eg Fig 5), inspired by Vinod Daniel's recounting at the 2001 Pest Odyssey meeting of instructions on solar heat disinfestation at the museum in Luang Prabang, Laos. The instructions had remained after a project by Bonnie Baskin (2001), who had consulted with CCI on solar disinfestation methods in order to find an eradication method suitable for use in a specific museum and amenable to religious concerns about the objects.

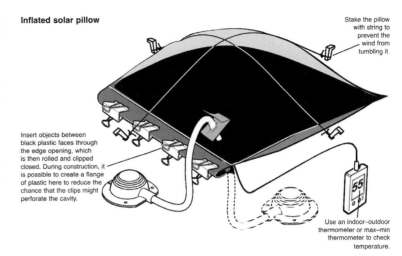

Inflated solar pillow

Stake the pillow with string to prevent the wind from tumbling it.

Insert objects between black plastic faces through the edge opening, which is then rolled and clipped closed. During construction, it is possible to create a flange of plastic here to reduce the chance that the clips might perforate the cavity.

Use an indoor–outdoor thermometer or max–min thermometer to check temperature.

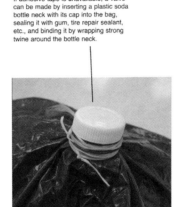

If adhesive tape is unavailable, a valve can be made by inserting a plastic soda bottle neck with its cap into the bag, sealing it with gum, tire repair sealant, etc., and binding it by wrapping strong twine around the bottle neck.

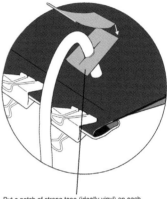

Put a patch of strong tape (ideally vinyl) on each clear cover. Make a small slit through the tape into just the clear covers. Insert the hose from an air pump or bellows and inflate both cavities until the black plastic faces are pressing together. Tape the slit closed with another patch of the same vinyl tape. Make a tab to grip the sealing tape by folding its end back.

Fig 5
Example disinfestation method.

Other methods now in use, such as controlled atmosphere fumigation, bagging with oxygen absorbers and so forth, are mentioned in the 'Respond' section and Bibliography. Key references are now generally available on the internet so we felt these did not need reproduction here.

Towards rapid visualisation of IPM concepts

A concern with instruction books is that a good portion of your primary audience can be lost if things aren't explained clearly enough. To address this, Tobunken supported the idea of developing a poster (Kigawa and Strang 2011) of the Technical Bulletin 29 IPM application guide (Fig 6).

Fig 6 (next page left)
IPM poster (front).

Fig 7 (next page right)
IPM poster (back).

文化財展示収蔵環境におけるIPMプログラム
「状況と対策の段階的モデル」

木川りか
独立行政法人国立文化財機構
東京文化財研究所

Tom Strang
Canadian Conservation Institute
Canadian Heritage

IPM（Integrated Pest Management：総合的害虫管理）は、博物館などの生物被害対策の方法として世界中で広く検討が進められている。本モデルでは、主に建物等の状況によってアつのレベルを想定し、一般的な対応と、それに対するIPMの活動の例を提示した。

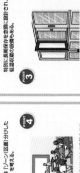

LEVEL 0

屋外環境である。

屋外収蔵庫、トーテムポール（風化が激しく、むきだし）など

改良案

POINT 1 周囲の下草、落ち葉、小枝などを払ってまわりに動物のすみかを減らす。

POINT 2 周囲の水はけをよくする。台座や地盤の段差を設け、地面への水の接触を減らす工夫をする。

POINT 3 日照、風雨により虫やカビなどによる劣化は、シートなどで覆うと発生しやすくなることもある。

POINT 4 カビ、虫害菌、ネズミなどの活動は、動物の食痕など定期的に調査する。

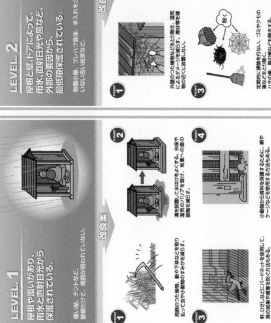

LEVEL 1

屋根や壁はあるが、雨水と直射日光から保留されている。

覆い屋、テントなど、屋根だけで、周囲が開放されている。

改良案

POINT 1 周囲の下草、落ち葉などを払ってまわりに動物のすみかを減らす。

POINT 2 遮水層などを設けるなどして、台座や温度のバリアを設け、地面からの湿気を減らす。

POINT 3 床、虫の網、ネズミなどを用いて、侵入の経路を設ける方法がある。

POINT 4 小動物などが資料を保護するために、網やケージなどを使用する方法もある。

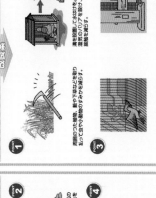

LEVEL 2

屋根と壁、トビラによって、雨水、直射日光が原因など、外部の要因から最低限保護されている。

物置小屋、プレハブ倉庫、手入れされていない古い建物など

改良案

POINT 1 外気などが資料にあたる場合、温度によるダメージがあるため、資料を暖かい物のそばにおく。

POINT 2 建物に雨が漏る場合など、屋根を修繕し資料を低湿かないようにしたりする。

POINT 3 定期的に清掃をすると、ゴミやカビ、虫などに注意しておく。ハチの巣、鳥の巣なども除去する。

POINT 4 資料に被害があった時や、建物内の資料の状況など、カビの原因となる要因の検討について資料を置く。

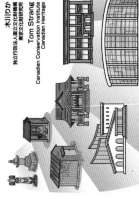

LEVEL 3

多くの歴史的建造物を想定した段階

換気は可能であるが、いわゆる空調設備等はなく、夏は暑く、冬は寒い。

建物内に手入れが行われている歴史的建造物。神社、民家、堀、蔵、寺社等の歴史的建造物。

改良案

POINT 1 建物に入る際に、虫を防ぐための網戸をつける。床のすきまや、壁などを0.3mm以下のすき間になるよう補修する。

POINT 2 入口付近に、虫の侵入を止止するための網戸をつける。

POINT 3 定期クリーニングのプランをたてて、1年に少なくとも数回掃除をする習慣をつける。

POINT 4 温湿度センサーなどを用いて、室内の温湿度を適宜に調査する。

POINT 5 少なくとも年に1回は収蔵庫内を隅々まで調査する。

LEVEL 4

博物館、美術館、文書館等の基本的な機能を果たしている。場合によっては、古い建物や、必ずしも博物館等専用に設計された建物ではない場合も含む。

改良案

POINT 1 新規に受け入れる資料の隔離室がある。

POINT 2 一時保管場所（燻蒸室、二酸化炭素処理のゾーン、低温処理ができるゾーン）を設置する。

POINT 3 年間クリーニングのプランを立てて、1年に数回、隅々まで清掃を行わないよう処理する。

POINT 4 文化財施設において収蔵・展示の環境調査や害虫の調査を行い、お客さまにも展示室の影響協力してもらう仕組みにしておく。

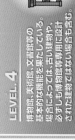

LEVEL 5

博物館、美術館、文書館等目的の建物が設計され、整備されている。

改良案

POINT 1 空調設備、カビの発生を防ぐための資料は、粉塵などが資料に及ばないよう設計される。

POINT 2 大きな施設では「ゾーン（区画）分けした IPMシステム」を考える。

POINT 3 ゾーンガードしたIPMシステムの導入

POINT 4 清掃スタッフをはじめとする関係者の合意をとりつつ、IPMの連携をとりないと生物被害の要因である埃や埃をへらすという認識を共有しておくらう。

LEVEL 6

建物の設計時点から資料の長期保存を念頭において、建物が特別に整備されている。

収蔵される資料の特性に応じて各々の収蔵庫に向けた設計がなされた建物。設計・環境段階から生物運動を防ぐためのあらゆるステップの検討がなされ、資料の保存に有効なステップが構築された建物。

改良案

POINT 1 特別に資料保存を目的に設計され、低湿収蔵の設備もある。

POINT 2 被害を受けやすい資料は、粉塵除去、低湿低温を行う一方の所、各々の専用の収蔵庫などで保管する。

POINT 3 被害を受けやすい資料は、個別収蔵を行う。害虫・カビなどの侵入しない設備に入れて保管する。

POINT 4 清掃スタッフやや警備員、レストランのスタッフなどに協力してもらえるようにしておき、館内の生物被害になる影響を受けないように連携をとっておく。

東京文化財研究所 保存修復科学センター

文化財展示収蔵環境におけるIPMプログラム
「状況と対策の段階的モデル」

木川 りか　独立行政法人国立文化財機構 東京文化財研究所
Tom Strang　Canadian Conservation Institute, Canadian Heritage

木川りか、Tom Strang：
文化財展示保存環境におけるIPMプログラム：状況と対策の段階的モデル、
文化財保存修復学会第27回大会、ポスター発表 (2005) を原案とする
〈参考〉
木川りか、Tom Strang：文化財展示収蔵環境におけるIPMプログラム：状況と対策の段階的モデル、『文化財保存修復学会誌』、49、132-144 (2005)
Tom Strang, Rika Kigawa: Levels of IPM Control: Matching Conditions to Performance and Effort. Collection Forum 21(1-2), 96-116. (2006)
Tom Strang, Rika Kigawa: Combating Pests of Cultural Property. CCI Technical Bulletin #29. Canadian Conservation Institute. (2009)

IPM (Integrated Pest Management: 総合的有害生物管理) は、博物館等での生物被害を予防し、抑制する方法として世界中で広く検討が進められている。ここでは、主に建物等の状況によって7つのレベルを想定し、一般的な状況と、それに対するIPMの活動の例を提示することを試みた。

＜レベル0＞
屋外環境である。

＜レベル1＞
屋根や覆いがあり、雨水と直射日光から保護されている。

＜レベル2＞
屋根と壁、ドアによって、雨水と直射日光や風などの外部の要因から、最低限保護されている。

＜レベル3＞
多くの歴史的建造物を想定した段階である。換気は可能であるが、いわゆる空調設備等はなく、夏は暑く、冬は寒い。

日光山 輪王寺 三仏堂

＜レベル4＞
博物館、美術館、文書館等の基本的な機能を果たしている。場合によっては、古い建物や、必ずしも博物館等専用に設計された建物ではない場合も含む。

＜レベル5＞
博物館、美術館、文書館を目的に建物が設計され、整備されている。

＜レベル6＞
建物の設計時から資料の長期保存を念頭において、特別に建物が整備されている。

Fig 8
Avoid, Block, Detect,
Respond – key messages on
the cover of Combatting
Pests of Cultural Property
(Strang and Kigawa 2009).

The Japanese-language poster was designed with quick to comprehend bold graphic cues for each level. We exhibited the draft at the CCI to elicit comments from people illiterate in Japanese to test comprehension of the graphic design. Suggestions were fed back for the final version. Graphic instructional posters are appealing for the simple reason that a picture of a broom or vacuum cleaner for example can be understood by people irrespective of the language(s) they speak.

Some ideas are not easily reproduced as graphics, and while many posters are one-sided, the Japanese poster on IPM incorporates more detailed technical matter on the verso (Fig 7).

These posters are designed for use throughout the nationwide facilities as part of Tobunken's collections-care education programmes and will aid staff with the student and volunteer activities which operate inside these facilities.

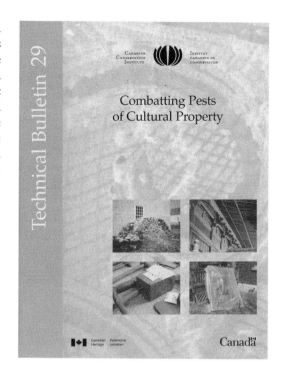

Envision ways to use the material

In addition to initiation and education, the book and poster can be used actively in surveys. The schemes can be marked up with circled features to get a snapshot of a system state, and adjacent features elicit thoughts on the next steps for improvement. This is in tune with overall risk reduction goals for the institutions.

Choosing an appropriate title and cover

The cover artwork (Fig 8) of *Combatting Pests of Cultural Property* (Strang and Kigawa 2009) clearly illustrates the four key elements of IPM – Avoid, Block, Detect and Respond. The use of the term 'combatting' in the title of the book (note Canadian spelling, Canadian Oxford Dictionary 2004) is intentionally militaristic. While Saul (1994) deflates 'manager' as 'drawn from the French word "ménager" or one who does domestic housework' and thus ideally suited for its use in IPM, Hartnack (1943) had a convincing rationale that the struggle against all pests reduces to one word – FIGHT.

Acknowledgements

We would like to thank the administrative support of the National Research Institute for Cultural Properties, Tokyo (Tobunken) and the Canadian Conservation Institute (CCI) and special thanks to Chie Sano, National Research Institute for Cultural Properties, Tokyo and Charlie Costain, Canadian Conservation Institute.

References

Agency for Cultural Affairs 2011 'Preservation and utilization of cultural properties', *in Administration of Cultural Affairs in Japan – Fiscal 2011*. Tokyo: Agency for Cultural Affairs. [In Japanese]

Åkerlund, M 1991 *Ängrar – finns dom …? Om skadeinsecter I museer och magasin*. Uppsala: Almqvist & Wiksell Tryckeri

Anon 2001 *A Cyclopedia of Japanese Museum Insects: To Facilitate Identification by Staff for Promotion of IPM in Museums*. Tokyo: IAI National Research Institute for Cultural Properties. [In Japanese]

Barber, K (ed) 2004 *The Canadian Oxford Dictionary*, 2nd edn. Don Mills, Ontario: Oxford University Press

Baskin, B 2001 'Solar bagging: Putting sunlight to work to eliminate insect infestations in mere hours'. *WAAC (Western Area for Art Conservation) Newsletter* **23** (1), 20–2 http://cool.conservation-us.org/

waac/wn/wn23/wn23-2/wn23-207.html
(accessed 19 August 2011)

Canadian Conservation Institute nd
Preservation Framework Online. www.cci-icc.gc.ca/crc/fw/index-eng.aspx (accessed
19 August 2011)

Canadian Conservation Institute nd *Ten Agents
of Deterioration*. www.cci-icc.gc.ca/crc/
articles/mcpm/index-eng.aspx (accessed
19 August 2011)

Canadian Museums Association 1997 *The
Official Directory of Canadian Museums and
Related Institutions, 1997–99*. Ottawa:
Canadian Museums Association

Doyle, A, Pinniger, D B and Ryder, S 2007
'Risk zones for IPM: From concept to
implementation'. *Collections Forum* **22**
(1–2): 22–31

Hartnack, H 1943 *Unbidden House Guests*.
Tacoma, Washington: Hartnack Publishing
Company

Kigawa, R, Nagaya, N, Sonoda, N, Hidaka, S
and Strang T 2003 'A review on integrated
pest management in museums and related
institutions: Basic concepts and practical
steps for introduction'. *Bunkazai Hozon-
Shuuhuku Gakkaishi* 47: 76–102. [In
Japanese, English abstract]

Kigawa, R and Strang, T J K 2005 Levels of
IPM. Poster presentation, *Annual Meeting of
the Japan Society for the Conservation of
Cultural Property (Bunkazai Hozon-
Shyuhuku Gakkai), Tokyo, 14–15 May 2005*

Kigawa, R and Strang, T J K 2011 *IPM
Programs for Various Environments:
Incremental Control Model*. Commercial
poster. Tokyo: National Research Institute
for Cultural Property

Michalski, S, MacDonald, M, Strang, T,
Tétreault, J and Williams, R S 1992 *A
Systematic Approach to the Conservation
(care) of Museum Collections – With
Technical Appendices*. Ottawa: Canadian
Conservation Institute

Ministry of Education, Culture, Sports,
Science and Technology (MEXT) 2008
Present Status of Museums in Japan. Tokyo:

MEXT www.mext.go.jp/component/a_
menu/education/detail/__icsFiles/
afieldfile/2009/04/27/1217880_3.pdf
(accessed 19 August 2011)

Pinniger, D B 2001 *Pest Management in
Museums, Archives and Historic Houses*.
London: Archetype Publications

Pinniger, D B 2008 *Pest Management: A
Practical Guide*. Cambridge: Collections
Trust

Standards Australia Joint Technical
Committee OB-007, Risk Management
2004 *Risk Management, AS/NZS
4360:2004*. Sydney: Standards Australia

Standards Australia Joint Technical
Committee OB-007, Risk Management
2004 *Risk Management Guidelines.
Companion to AS/NZS 4360:2004*. Sydney:
Standards Australia

Saul, J R 1994 *The Doubter's Companion. A
Dictionary of Aggressive Common Sense*.
Toronto: Viking Penguin Canada

Strang, T J K 1998 'Another brick in the
wall', *in* Åkerlund, M, Bergh, J-E,
Stenmark, A and Wallenborg, I (eds) *The
3rd Nordic Symposium on Insect Pest
Control in Museums*. Stockholm: PRE-MAL
(Pest Research and Education – Museums
Archives and Libraries) and ICOM
(International Council of Museums),
10–29

Strang, T J K 1999 'A healthy dose of the past
– a future direction in herbarium pest
control?' *in* Metsger, D A and Byers, S C
(eds) *Managing the Modern Herbarium: An
Interdisciplinary Approach*. Vancouver:
Elton-Wolf, chapter 3, 59–80.

Strang, T J K and Kigawa, R 2006 'Levels of
IPM control: Matching conditions to
performance and effort'. *Collection Forum*
21 (1–2): 96–116

Strang, T J K and Kigawa, R 2009
'Combatting pests of cultural property'. *CCI
Technical Bulletin* 29. Ottawa: Canadian
Conservation Institute www.cci-icc.gc.ca/
crc/articles/mcpm/chap06-eng.aspx
(accessed 19 August 2011)

4

Integrating IPM risk zones and environmental monitoring at the Museum of London

Adrian Doyle*, Chloe Evans† and Maria Yanez Lopez‡

*Museum of London, London, UK, adoyle@museumoflondon.org.uk (author for correspondence)
†Student on placement at Museum of London, chloenevans1@aol.com
‡Intern at the Museum of London funded by The Clothworkers' Livery Company, London,
myanezlopez@museumoflondon.org.uk
www.museumoflondon.org.uk/

ABSTRACT

The Museum of London (MOL) has introduced an integrated pest management (IPM) risk zones concept which uses electronic colour-coded gallery and store floor plans mapped and stamped with IPM risk zones and insect trap location symbols. These have been incorporated into the Hanwell environmental data management system and, together with a colour-coded trapping spreadsheet, make it easy to relate the environmental data to pest risk, which has helped determine the appropriate level of response to infestation or environmental problems.

KEYWORDS

Risk zone protocols, indicator species, floor plans, trapping spreadsheet, Hanwell environmental monitoring system

Introduction

The relationship between insect pest populations and environments has been recognised for some time and can be useful to contradict or correlate pest populations with environmental problems, hopefully mitigating significant infestations (Pinniger and Winsor 1998).

Particular pests can act as indicator species which are specific to environmental problems. In particular, webbing clothes moth (*Tineola bisselliella*) activity is more widespread with increased temperature, while silverfish (*Lepisma saccharina*) activity is directly related to damp or water ingress.

In common with other organisations, the Museum of London (MOL) records the collated insect trapping data on spreadsheets, while environmental data is managed separately using Hanwell radiotelemetry data loggers that provide a graphical display.

Formerly, these two processes have acted independently of each other, but by cross-referencing IPM data with environmental data, specifically relating to IPM risk zones (Doyle *et al* 2007) with dedicated protocols, the bigger picture of insect populations, their movements and their relationship with the environmental conditions can be more fully understood.

Introducing risk zones and pest prevention protocols

The IPM risk zones concept provides a framework for implementing the appropriate level of control necessary for minimising the chances of infestation or pest damage. According to the vulnerability of the collection, from high risk to low risk, specific protocols are in place to ensure best practice standards.

The highest priority for collections care conservators, and where resources can be best targeted, are stores holding high-risk material, specifically those that are highly attractive to pests; these are designated red zones.

Red zones

In these spaces, the following mandatory rules are applied as a minimum standard:

- temperature and relative humidity levels are continuously monitored and recorded through the telemetric system Hanwell Radiolog
- blunder traps and moth pheromone traps are checked monthly
- floor plans showing the designated risk zone and trap locations are installed
- no food is allowed – 'water only to drink' signage is in place
- doors leading to and from the store have bristle seal strips fitted
- ceiling lights are sealed units or do not have diffusers
- a minimum of one deep clean per year is carried out.

Only objects that are visually checked to be free of pests are allowed into these stores, but it should be noted that this is difficult to enforce and visual inspections for pests are not necessarily a 100 per cent guarantee.

Additionally, a member of the collections care staff carries out a weekly store check to ensure that the areas are kept tidy and up to the designated standard for that store.

Orange zones

Display areas with high-risk collections are designated orange zones. These areas benefit from high maintenance standards as they are galleries open to the public on a daily basis,

but equally environmental management and gallery cleaning protocols contribute significantly to pest prevention. It is worth noting that gallery staff (visitor hosts) look after these spaces daily and prevent people from eating and drinking outside designated areas. This policing, combined with monthly gallery team maintenance inspections, helps to anticipate problems and enable actual problems to be addressed more quickly.

As part of our standard environmental monitoring policy, the MOL monitors the environment in all galleries, and insect blunder and moth pheromone traps are checked monthly. It is standard practice, as part of our display criteria, that high-risk collections are displayed in good quality display cases wherever possible, and traps (which can be disguised to blend in more successfully with the display environment) act as an additional level of security.

The MOL building has an open plan gallery layout, with no separated or enclosed gallery spaces. Therefore, the whole of the display space on two floors is in fact one display area. This is designated an orange risk zone in spite of the lower risk rating of the materials in some gallery spaces. Treating the entire display area as one gallery enables us to take a museum-wide approach to the IPM strategy. In essence, the procedures and protocols have to remain of high standard throughout in order to protect the most vulnerable collections.

Yellow and green zones

The remainder of the MOL building, including isolated areas containing objects which are less vulnerable to pest attack, are designated low-risk yellow zones. Non-collection areas are designated no-risk green zones.

In terms of pest prevention, maintaining high standards of housekeeping is still deemed crucial in these spaces, particularly when they are near high-risk areas. Where food preparation or consumption is permitted, cleanliness is essential as food waste can attract rodents and insects that can cause infestation problems. Dead rodents can become a food source for moths and other insects, and these could easily spread into high-risk collections areas, thereby defeating the holistic approach to IPM.

Fig 1
Symbols for insect blunder trap, webbing clothes moth, high (A) risk zone, Hanwell data logger and bristle/brush strip.

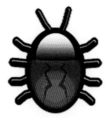

Insect blunder trap symbol

Webbing clothes moth symbol

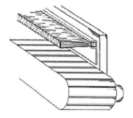

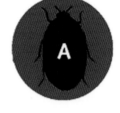

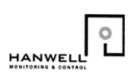

Bristle/brush strip symbol

High (A) risk zone symbol

Hanwell data logger symbol

Combining the information on floor plans on the Hanwell Radiolog system

The initial stage in this process was to gather electronic copies of floor plans of the entire museum building, including all stores, gallery spaces and office. The floor plans give an instant visual guide to the different spaces and so the relationship between areas becomes more apparent. Copies of the floor plans were provided by the Facilities department, which uses them for building maintenance purposes.

Fig 2 (below)
An example of risk zones, Hanwell data logger and pest trap locations plotted on gallery floor plan.

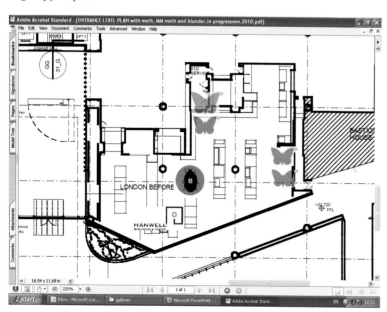

The plans were manipulated using Adobe Photoshop software to remove unnecessary detail such as irrelevant text and grid lines, and to add in any features such as temporary walls. The software also allows us to update the floor plans in line with any developments within the museum. The plans were colour-coded according to the agreed risk zone in order to immediately alert users to the specific pest protocols for that area.

The next stage was to use additional software, Adobe Acrobat Pro, to mark the locations of the following on each floor plan:

- Hanwell data loggers
- blunder traps
- pheromone moth traps (both Indian meal moth and webbing clothes moth)
- bristle/brush seals.

A specific symbol was chosen to represent each of the above, and these were added to the gallery plans using the software's stamp tool. An insect symbol was chosen for the blunder trap and a butterfly symbol for the moth traps (Fig 1). It is a simple and effective way to accurately mark trap locations since each stamp can be dragged into position with the option of repositioning them when the traps are moved. This is particularly useful for gallery spaces where numerous traps are installed as it acts as a simple map of trap locations, thereby helping to speed up monitoring (Fig 2).

Using Adobe Acrobat editing software, by right-clicking each symbol, a comment text box can be created where further information can be recorded, such as the trap number, catch data and any special instructions. In normal use, with Adobe Reader, by rolling the computer mouse cursor over the icon, this information is displayed in a pop-up box and does not require the editing software.

With this easy-to-use system, data is simple to access and interpret. In particular, trapping data can be displayed within the maps and made available to other staff. For example, at the MOL Docklands this proved to be very helpful when working with external pest control contractors. We were able to email floor plans alongside the trapping data, so helping them to assess the extent and location of the moth activity and develop appropriate treatment proposals. Additionally, feedback from the contractors about using this technique was favourable.

For each storage area, the relevant plan, stamped with the appropriate symbols, was printed and displayed on the wall (Fig 3). This ensures that when staff enter the space they are immediately aware of the pest risk, and are able to locate traps and the environmental logger.

We also use spreadsheets to record monthly trapping data, which enables us to analyse the data and provides graphical representation that can be used in the annual report for each location. Each museum area identified on the floor plans is recorded on the spreadsheet and these are colour-coded with the appropriate risk zone. In this way, trapping data is immediately related to the risk to the collections and therefore helps prioritise our preventive conservation activities.

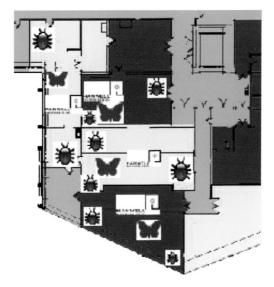

Fig 3
An example of risk zones, Hanwell data logger and pest trap locations plotted on the store floor plan. See Figure 1 for an explanation of the symbols.

Inputting floor maps into Hanwell environmental monitoring system

The next development for this project was to combine these risk-zoned floor plans with the Hanwell environmental management software. Hanwell software has the option for inputting floor plans on which Hanwell sensor icons can be placed. This is useful for easily identifying the location of the loggers that alert staff to low or high environmental conditions as well as battery and transmitter problems. Store risk-zoned (Fig 4) and gallery trap symbol-stamped floor plans were installed into the Hanwell system (Fig 5). This has proven to be particularly effective when correlating environmental problems with pest-related issues as the relationship between the two can be clearly seen.

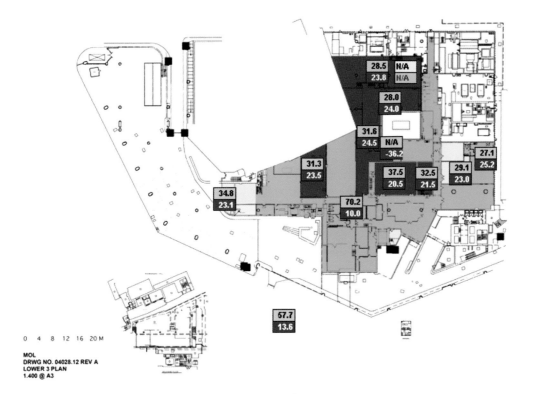

0 4 8 12 16 20 M

MOL
DRWG NO. 04028.12 REV A
LOWER 3 PLAN
1.400 @ A3

Fig 4
Hanwell floor plans with risk zones for stores.

Fig 5
Hanwell floor plans with stamped symbols for galleries.

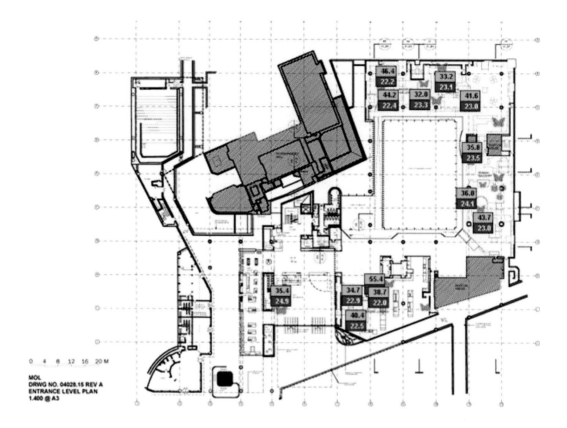

Certain problem areas were identified as contradicting our initial decisions and preconceptions. For example, in a particular high-risk store, trapping data showed significant numbers of silverfish, which are an indicator species for damp conditions; while the Hanwell logger showed the relative humidity and temperature were well within normal parameters for that store. Further investigation of the area revealed a faulty water pipe that was leaking near to the insect trap (the floor was wet). This confirmed that it was a localised problem and not a store-wide environmental issue.

Conclusion

As a result of these initiatives, we have been able to identify areas at risk more effectively and MOL staff have been able to identify priorities more quickly. We can now integrate the environmental data with the pest trapping data by cross-referencing the results from the monthly trapping data with the monthly environmental records. This enables us to quickly identify areas of concern and react appropriately. Only through analysing environmental conditions and pest activity together can these issues be fully understood and used to prevent pest activity in the future.

Acknowledgements

The authors would like to acknowledge the assistance, encouragement and technical support from Sharon Robinson, Collections Care Manager and Tanya Pollard, HUB Collections Care Development Officer, both of the Museum of London.

References

Doyle, A M, Pinniger, D and Ryder, S 2007 'Risk zones for IPM: From concept to implementation'. *Collection Forum* **22**(1–2), 23–32

Pinniger, D B and Winsor, P 1998 *Integrated Pest Management: A Guide for Museums.* London: Museums & Galleries Commission

5

Mapping museum pest activity: a review of the development of KE EMu as a tool for pest management

David A. Smith

Natural History Museum, London, UK, D.A.Smith@nhm.ac.uk
www.nhm.ac.uk

ABSTRACT

The vision of a collaborative initiative, led by the Natural History Museum, London (NHM) to develop existing functionality within KE Software's collections management system (KE EMu) to capture and visualise pest management data has now been realised. Emulating some of the functionality from professional geographical information system (GIS) software, but within a familiar user interface, the system displays pest trapping data in a spatial context on building plans.

   Collections management staff will finally have the capability to relate pest hot-spots to high-risk collections areas, observe seasonal patterns and check the effectiveness of any pest prevention measures.

KEYWORDS

Pest management, KE EMu, spatial mapping, KE Software, museum

Historical background

The use of geographical information system (GIS) technology as a tool to manage the risk of damage by insects has long been established (Roberts *et al* 1993). With increased availability of the microcomputer, the early 1990s saw an increase in computer-aided mapping to assist in the protection of agricultural crops. Under pressure to minimise the risk to human health and environmental quality from pesticide use, farmers were turning to the science, now called 'precision agriculture', to maximise their returns. By the late 1990s, the application of spatial technology as applied to pest management, both indoors and outdoors, appeared in many papers published in entomological journals (Brenner *et al* 1998; Arbogast *et al* 2000; Stafford 2000).

Application of GIS to museum pest management

Understanding the variety of factors that encourage or deter insect pest populations is critical to the development of an effective pest management programme. The advantage of GIS, in achieving this, is the ability to plot temporal and spatial data, and use visualisation tools to observe trends and relationships that would not necessarily be apparent using more traditional methods.

   Spatial mapping would, therefore, seemingly appear to be an incredibly useful tool in the effort to understand the dynamics of pest populations within museum buildings. Indeed, a small number of institutes have experimented with the technology (Monk *et al* 2002; Bryant 2006); however, the scarcity of published articles on this subject is very telling. GIS is a very powerful database system geared towards the professional user that requires considerable and expensive training in order to produce a proficient user. As there are now a number of freely available GIS mapping applications, it is reasonable to assume that it is not necessarily the cost that is inhibiting its use in the museum sector, but the complexity that is perhaps the barrier.

Adaptation of KE EMu

KE Software's electronic museum documentation system (KE EMu) is a comprehensive tool which facilitates a museum's primary obligation – the duty of care for its collections. Within an extensive array of collections management modules, core processes are documented and cross-referenced to the object thereby maintaining up-to-date information about its status and preserving a history of activity and transactions. With over 2,500 users across the globe (Sullivan 2009, pers comm), it is the software of choice for over 100 institutions and governmental bodies, large and small, with collections ranging from natural history, art and culture, library and social history.

In 2009 a consortium of four large international museums, led by the Natural History Museum, London (NHM), commissioned KE Software to extend their collections management software to include the capability to capture pest monitoring data. This would be a generic development that could be deployed to all institutes that already use KE EMu. However, in order to be as compatible as possible for future data interchange with other non-EMu databases, at the heart of the development would be the closest thing to a universal data standard for pest monitoring databases – the 'Suggested field list for pest observation databases' produced by the Integrated Pest Management Working Group (IPM WG) (2009), an influential international group set up to promote and facilitate good practices in pest management.

Once pest observations had been captured, there would be a series of tools to allow collections staff to interrogate the data and produce clear and informative reports to aid decision making. There would be the familiar pivot-table Excel reports providing users with summary tables and charts. However, the real vision of the consortium was to extend the KE EMu mapping web-service, beyond its regular use of creating interactive maps of the physical location of specimens on display, to emulate a GIS interface in which the spatial distribution of pest data could be observed and analysed. This would be the first step of a much grander vision to analyse the dynamics of insect pest populations in conjunction with the environmental conditions within a building.

A review of the system

This review is based upon the first incarnation of the development prior to installation at any institution and is, therefore, a work in progress. Since this review was written, a number of institutes have tested it and contributed recommendations for refinement and improvements to be included in future versions.

Data capture

With the familiar arrangement of tabbed windows within each of the modules, populating the data could not be easier for those who use KE EMu in their everyday activities.

Pest trapping data consists essentially of two components:

- data which describes the physical location of the trap
- data which records the diversity and frequency of insects found at those traps.

Within KE EMu, this data actually spreads across three data tables (Fig 1). The existing locations module is used to describe the physical location of each of the traps within the building. The data is arranged hierarchically so that it is possible to query locations within increasingly smaller zones, for example, from site to building to floor to room to cabinet levels. Of paramount importance within the location module are the x–y–z fields that record the co-ordinates of the traps' exact locations on the plans of the institute within the mapping interface (see IPM viewer).

The trap record provides the facility to describe the physical location of the trap in more detail and in terms of proximity to risk factors that may affect the pest numbers, for example, distance from an open-able window (Fig 2). Since each trap has a unique location,

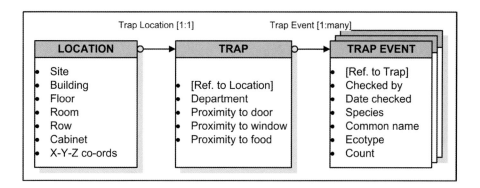

Fig 1
Pest management data is split across three modules.

there is a one-to-one relationship between the trap and location record. Once these have been created for all the traps in the array they become static, with all new data being created in the trap event module.

A trap event record is spawned from the trap record of interest, using a menu option, and it is automatically linked to it (Fig 3). The trap event record can then be completed with the information pertaining to the diversity and count of insect species found at that trap on a specific date. Entry of common name, species and ecotype is from a fixed list of lookup values. As all three are linked, entering either a common name or species value will automatically populate the other two columns. An additional useful feature that KE Software has introduced is a summarised view of the history of trap events relating to each trap (*see* Fig 2).

Query and reports

A combination of any of the fields in any of the three modules can be queried, making for powerful multi-faceted analysis. At the simplest end of the spectrum, the trap events can be queried for a specific pest insect species, or a number of species with a chosen range of counts, to identify which traps are species hotspots. This query can then be refined to identify which of these traps are in close proximity to windows and doors and, of those, which occur on a specific floor of the building or collection area.

Since the pest data is one component within a much larger database system, the results of a trap event query can then be cross queried with specimen information to assess whether any specimens that are susceptible to insect damage are in the same vicinity as those traps with recurrent high-risk insect occurrences.

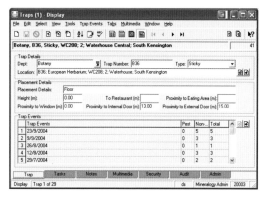

Fig 2
The trap record captures placement information. A history of trap events, associated with the trap, dynamically appear in the grid at the bottom.

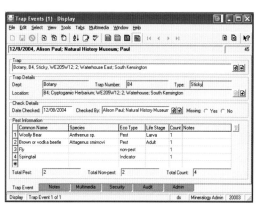

Fig 3
All new insects found in a trap are recorded in a trap event record.

Once a useful sub-set of data has been retrieved, there are a number of ways to explore the results, but this review shall look at the following:

• output to one of the 15 pre-defined pivot table reports

• output to the IPM viewer, which displays the data at each trap as a segmented pie chart on the plans of the institute.

Pivot tables and charts

KE Software has provided many pre-configured options that enable one to dice and slice the data (Fig 4). However, if one is familiar

Fig 4 (right)
The IPM modules come with
over 15 report templates.

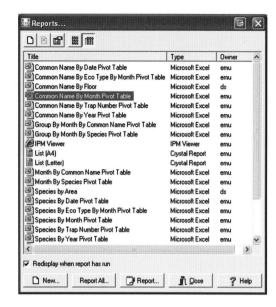

Fig 5 (below, top)
A graph showing the
number of insect pest species
found in traps during 2004.

Fig 6 (below, bottom)
A graph showing the
number of insect pest species
found in adjacent enclosed
spaces during 2004. There
are clear differences in the
diversity between zones.

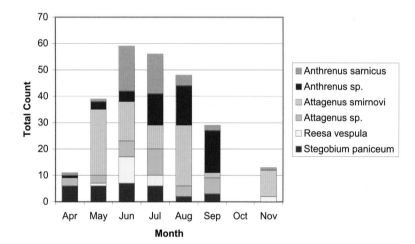

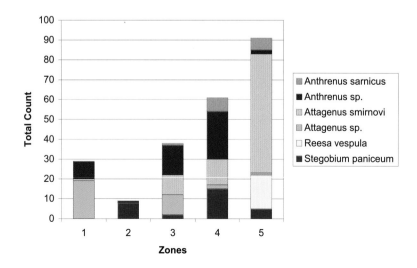

with the process of importing data into MS Excel, it is possible to construct alternative pivot table reports for oneself, although there are very few combinations which have not been looked at.

Selecting to report on species by month for any one particular year clearly shows peaks and troughs in insects being caught in traps (Fig 5). Insects that are of little interest can be filtered out using the options associated with the chart legend. Slicing the same data to look at the distribution of those insects within enclosed spaces across the trap array gives a good idea of differing insect activity (Fig 6). One can also begin to see differences in the range of species being caught in each zone. But it is the combination of both of these which provides specific information about where the hot-spots are and at what time of the year. Generating these charts for the same period as each trap event (for example, monthly), and comparing them side by side, may well reveal some recurrent patterns or anomalies, but it is not possible to put this into the context of the building. This is where the plotting of data on a map comes into its own.

The IPM viewer

The IPM viewer is built from the open source JavaScript application called OpenLayers. It integrates Google charts to be able to plot the trap event data as segmented pie charts. When an IPM viewer report is run, a mapping interface is opened in the default web browser. After a brief pause, as the application loads the data, the navigation tools appear and the data for each trap is summarised in segmented pie charts, where the diameter is proportional to the count. The location of the graphic on the map is defined by the x–y–z co-ordinates as recorded in the location module.

The key elements of the IPM viewer (Fig 7) are as follows:

- **Map display window**

    Area in which the selected floor plans and pie charts are displayed. Zoom and pan functionality is provided by buttons in the top left, or by the mouse wheel and click-n-drag, respectively. Placing the cursor over a marker initiates a pop-up box in which the top four entries are displayed.

- **Legend**

  To the right-hand side of the map window, the legend provides a key to the pie chart diameter and the segment colours. The range of counts pertaining to each of the diameters is calculated at the point of data load using an algorithm based on the total insect count of all the trap event records being reported on. This means that the scale is fixed within a single session, allowing for comparisons to be made as the data is filtered. Likewise, the colours are allocated to each of the variables at the point of data load to allow for data comparison within each session.

- **Views**

  Floor plans for each of the levels within an institute can be selected from the drop down menu. In brackets, adjacent to each of the entries, is the total count of insects relating to the traps on that level. Selecting a floor level filters the trap event data to display only that which relates to traps on that floor level.

- **Category**

  Radio buttons permit the data to be viewed in terms of number of species, common name, or numbers of each ecotype. Ecotype consists of generic values that group species according to whether they are pests, non-pests or indicators of humidity.

- **Filter dropdowns**

  Depending on the category selection, the entries in the adjacent drop down change dynamically. Single or multiple values can be selected to filter the display to only the insect, or insects, of a chosen life stage. These can be further filtered by selecting single or multiple trapping dates.

- **Date sequence player**

  Having selected the insect criteria of interest and chosen a range of dates, the date sequence player will commence a time-lapse sequence that displays the pie charts relating to the data for each of those dates. This clever little tool allows observations to be made about changes in pest distribution over time.

  Loading into the IPM viewer the same subset of data as that charted earlier very clearly highlights the location of the insect hot-spots (Fig 8). The suggestion of different insect sub-

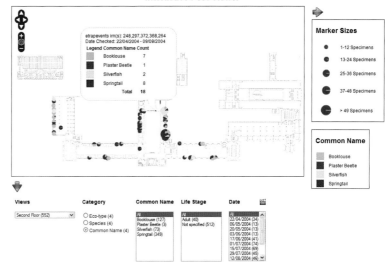

communities within different areas is now more clearly seen.

Discussion

Familiarity with the interface

The inclusion of pest monitoring data capture within the KE EMu collections management system really enhances it as a documentation toolkit to facilitate the effective management of collections. The addition of these two new modules negates the need for multiple databases. Staff, who already use KE EMu in their

Fig 7
The IPM viewer web interface displaying the spatial distribution of indicator species. See the text for a description of the key elements.

Fig 8
The spatial arrangement of insect diversity within adjacent enclosed spaces is clarified using the IPM viewer (see Fig 6).

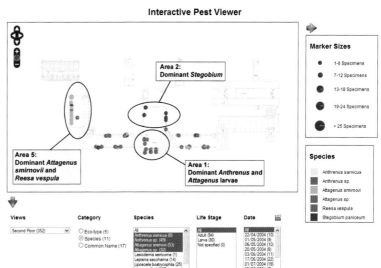

everyday working practice, will need no further training to interact with these two new modules; data entry, record navigation and query methods will all be familiar.

Proximity to collections information

The purpose of any pest monitoring programme is to protect the collections from insect damage through a strategy of risk management based upon the observed pest population. Thus, integrating the primary pest data with collections information in the same system will allow for more refined analysis, of which collection units are at greatest risk. When collections management resources are already stretched thinly and finances are increasingly hard to come by, prioritised preventive measures are what are called for.

Institution-wide analysis

Often it is individuals who have a sense of the actual risk to a collection unit, by virtue of them regularly checking the traps or responding to pest infestations in their area of responsibility. Although regular reports may be submitted to IPM programme leaders, producing effective strategies and long-term plans based on disjointed information is not an easy task. The IPM viewer breaks down these invisible department-centric boundaries and permits the user to observe the insect population in adjacent areas and make interpretations about the influence of insects in adjacent spaces – perhaps making observations about harbourage and migration pathways between life stages.

The viewer really succeeds in providing an institution-wide perspective on the pest situation. Through a colour-coded system of pie charts, whose diameters reflect the insect counts at each trap, the eye can very quickly identify hot-spots and delineate zones of differing insect 'species communities'. This is critical for developing over-arching strategies and targeting limited resources to high priority needs. It also facilitates dialogue with senior staff who may know little about insect pest management, but who make decisions on how budgets should be allocated. A well placed spatial map of high-risk insect species may well sway important decisions.

Compatibility/accessibility

At the heart of the development of the KE EMu pest management modules are the recommendations of the influential IPM WG. By adhering to these guidelines it is anticipated that data exchange or cross-platform querying may be achievable. Thus a more strategic nationwide, or even global, view of pest populations, and their distribution within cultural institutes and historic buildings, could be possible in the future.

Room for improvement

It would not be a thorough review if any weaknesses were not highlighted. Fortunately, there are very few to speak about and, as the prototype is rolled out, some of these will no longer be an issue.

As far as the layout of the data capture screens go there is nothing negative to report. However, in terms of populating those screens, it was discovered that creating trap event records one at a time could become quite tiresome, particularly where there are large trap arrays or where the majority of the traps are insect-free. Efficiency of data creation may well be improved by providing a mechanism to batch populate new trap event records with recurrent information, such as the trapper's name and date checked, leaving the user to enhance the records with insect information as appropriate.

When a viewer report is run on a queried sub-set of data, the colours that make up the legend are allocated to the unique 'species' and 'common name' values within that data-set, and this association is fixed for the duration of the session. A crucial mistake is that this association is not permanent. This means that, whilst comparison can be clearly made between any filtered view of the same data within one report, they cannot be compared with views of data from any other query, since the colours will have been reassigned to different values. The strength of the IPM viewer is its use of colour to identify spatial trends and temporal patterns. If the key colours are not uniquely assigned to the data values then it becomes impossible to view these trends.

The viewer is not perfect, but then this is a prototype. KE Software is aware of the points raised above and is in discussion as to how to improve the viewer for future versions. It is likely that these concerns will be resolved before this paper comes to press.

Conclusions

The vision of the IPM consortium, together with the technical expertise of KE Software delivering that vision, has really extended the capability of KE EMu as a tool to more effectively manage the collections. It will provide museums, as guardians of cultural heritage, with additional information to define policy and procedure and implement strategic plans to ensure the longevity of the collections.

There is huge potential in this viewer to aid pest management programmes. For example, future developments could provide the facility to overlay the location of certain objects (like those on open display). In addition, enabling proximity searches between layers could highlight those specimens most susceptible to insect damage occurring within a specified radius of an insect hot-spot. These enhancements would permit a more proactive and targeted approach to pest management in the future. There are exciting times ahead.

Acknowledgement

I would like to extend my gratitude to the staff at KE Software for facilitating the installation of this prototype at such short notice.

References

Arbogast, R T, Kendra, P E, Mankin, R W and McGovern, J E 2000 'Monitoring insect pests in retail stores by trapping and spatial analysis'. *Journal of Economic Entomology* **93**(5), 1531–42

Brenner, R J, Focks, D A, Arbogast, R T, Weaver, D K and Shuman, D 1998 'Practical use of spatial analysis in precision targeting for integrated pest management'. *American Entomologist* **44**, 79–101

Bryant, J A and Kang, J 2006 'A GIS tool for the interpretation of pest monitoring data in a museum environment'. Poster presentation at the *21st Annual Meeting of the Society for the Preservation of Natural History Collections, Albuquerque, 23–27 May 2006*. www.museumpests.net/resources/presentations/RMM%20IPMGIS%20Analysis%20poster.jpg (accessed 15 February 2011)

Integrated Pest Management Working Group (IPM WG) 2009 *Suggested Field List for Pest Observation Databases*. www.museumpests.net/resources//Updated%20_Field_List_for_Pest_Observation_Databases.pdf (accessed 15 February 2011)

Monk, R R, Butler, J B, Sandoval, R V and Dawson, E M P 2002 'GIS as a method for improving IPM'. Presentation at *17th Annual Meeting of the Society for the Preservation of Natural History Collections, Montreal, 8–13 May 2002*.www.nsrl.ttu.edu/about/Presentations_Posters/projects/spnhc2002/GIS_files/frame.htm (accessed 15 February 2011)

Roberts, E A, Ravlin, F W and Fleischer, S J 1993 'Spatial data representation for integrated pest management programs'. *American Entomologist* **39**, 92–107

Stafford, J V 2000 'Implementing precision agriculture in the 21st century'. *Journal of Agricultural Engineering Research* **76**(3), 267–75

Sullivan, B, KE Software, pers comm, June 2009

6

Developing and implementing an integrated pest management concept in the large collections of the National Museums in Berlin

Pascal Querner* and Stephan Simon†

*University of Natural Resources and Life Sciences, Vienna, Austria,
pascal.querner@boku.ac.at (author for correspondence)
†Rathgen Research Laboratory, National Museums Berlin, Berlin, Germany

ABSTRACT

The National Museums of Berlin form part of the Prussian Cultural Heritage Foundation. The sixteen museums comprise one of the largest collections internationally, in which pests have widely infested and damaged collections. For example, in the Ethnological Museum different pest species, mainly webbing clothes moths, *Tineola besselliella*, biscuit beetles, *Stegobium paniceum*, common furniture beetles, *Anobium punctatum,* and tobacco beetles, *Lasioderma serricorne,* have been harming the collections for many years. In the past, the museums had a treatment chamber where Lindane (*gamma*-hexachlorocyclohexane) was used for pest control. In 2002 a nitrogen tent and a freezing chamber were installed in the Ethnological Museum, which are used mainly for their objects but also by other institutions of the Foundation. Integrated pest management (IPM) is an essential component of preventive conservation, but only a few museums, until recently, had set up a monitoring programme. However, because of a large clothes moth outbreak in 2008, and repeated infestations in the collections of the Ethnological Museum, a full-time one-year IPM post was created and based in the Rathgen Research Laboratory. The aim was to: implement an IPM policy for the ethnological collection and for other museums within the Foundation, train staff, prepare an insect pest species collection, and engage in research and dissemination of information. This paper reports on the difficulties, results and milestones from this first experience with IPM at the National Museums in Berlin, and discusses the question of what size a museum or collection should be to justify employing a full-time or part-time IPM specialist.

KEYWORDS

Integrated pest management, IPM, strategy, raising awareness, staffing level

Introduction

Integrated pest management (IPM) was introduced into museums in the 1980s and 1990s (Story 1986; Linnie 1987; Albert and Albert 1988). Today, part of preventive conservation is focusing on the prevention of pest infestations and the reduction of pesticide application. This is achieved by sealing the building against pests (Fig 1), adapting the climate, regular cleaning and good housekeeping, quarantine, monitoring pest infestations with traps and, only when necessary, using non-harmful methods to treat the infested objects (Linnie 1996; Åkerlund *et al* 1998; Jessup 1998; Strang 1999; Kingsley *et al* 2001). More recently, several publications have described the implementation of IPM programmes and have helped to establish the concept of IPM (Boylan 2004; Pinniger 2004; Pinniger and Winsor

2004; Strang and Kigawa 2006; Brokerhof *et al* 2007a; Brokerhof *et al* 2007b; Pinniger 2008). Treatment methods have been changing from the use of chemicals (pesticides or fungicides) to chemical-free methods such as heating, freezing or anoxia treatments, mainly using nitrogen or carbon dioxide (Gilberg 1989, 1991; Strang 1992; Maekawa and Elert 2003; Pinniger 2003). Chemical pesticides are only used in emergencies or when no other method can be applied; for example, if only a few days are left before the opening of an exhibition. Insect pest monitoring is an important part of IPM which involves detecting pests, identifying them correctly, and locating infested objects or problems resulting from the building (Gilberg and Roach 1991; Child and Pinniger 1993; Cox *et al* 1996; Ackery *et al* 1999; Pinniger *et al* 2004). Today, most museums in German-speaking countries (Germany, Austria or Switzerland) reject the use of pesticides and have started implementing an IPM programme or at least parts of one. However, few museums have a long-term monitoring protocol in place and IPM is not yet standard for all large collections (Reichmuth *et al* 1991; Reichmuth *et al* 1994; Unger *et al* 1996; Brand and Wudtke 1997; Pinniger 1998; Ranacher 1998; Wießmann 1998; Unger 2002; Wudtke 1996, 2002, 2003; Noldt and Michels 2007; Querner and Morelli 2010a,b).

The National Museums of Berlin form part of the Prussian Cultural Heritage Foundation, which was set up in 1957. The Foundation is one of the world's major cultural organisations, with the State Museums of Berlin, the State Library, the Secret State Archives, the Ibero-American Institute and the State Institute for Music Research closely linked to form a network for cultural dissemination. The State Museums of Berlin comprises sixteen individual museums devoted to the ancient cultures of the Mediterranean (Egypt, the Near East, Greece, Rome), to all branches of modern art and culture (painting, sculpture, graphic art, applied art, photography, film, video), to the great non-European cultures (India, the Islamic world, East Asia), and to the ethnology of all five continents. Altogether, the State Museums of Berlin form the largest and most multi-faceted museum complex in Germany and one of the largest in the world.

Until recently, only a few museums in Berlin had an IPM monitoring programme in place. Following the extensive outbreak of webbing

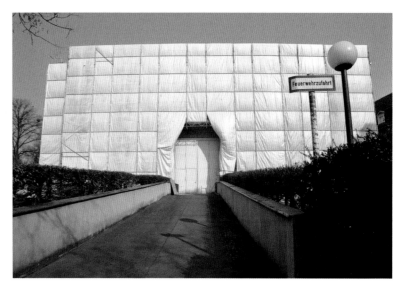

clothes moth in 2008 and the repeated infestations in the collections of the Ethnological Museum, and after discussions with the head of the Foundation, museum staff from the Ethnological Museum who were concerned with preventive conservation, and the Rathgen Research Laboratory, established a full-time IPM position. The aim was for the post-holder to: investigate a sustainable and long-term solution for the whole Foundation, implement an IPM concept in the ethnological collection and in other museums within the Foundation, train staff, prepare an insect pest species reference collection, and engage in research and dissemination. The costs for the one-year full-time position would be the equivalent of a large-scale chemical insecticide treatment (using sulfuryl fluoride) for the infested storage building of the Ethnological Museum.

Material and methods

The IPM position was based in the Rathgen Research Laboratory in Berlin for 12 months (October 2009 until October 2010). During this time an insect pest monitoring programme was introduced into a number of museums, including: Ethnological Museum, Museum of European Cultures, Bode Museum, Museum of Islamic Art, Egyptian Museum, Kunstbibliothek (Art Library), Museum of Photography, Kupferstichkabinett (Museum of Prints and Drawings), Museum of Musical Instruments and Geheimes Staatsarchiv (Secret State Archives Prussian Cultural Heritage Foundation). Collections such as those in the Ethnological Museum and

Fig 1
Outside renovation of the storage depot to seal the building better,
Ethnological Museum Berlin.
(Photo © Pascal Querner)

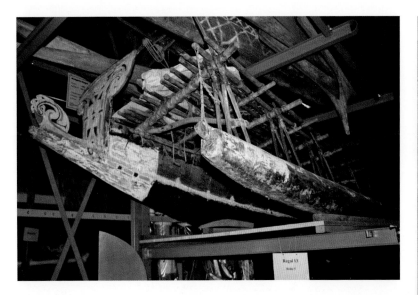

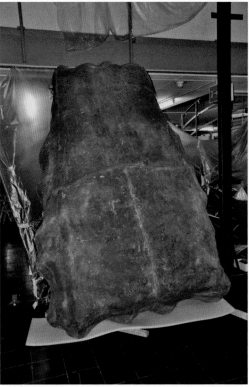

Fig 2 (above)
Infested wooden boat from Pacific islands in the storage depot of the Ethnological Museum Berlin.
(Photo © Pascal Querner)

Fig 3 (far right)
Infested Tibetan boat in the storage depot of the Ethnological Museum Berlin.
(Photo © Pascal Querner)

Fig 4 (below)
Infested Vodou temporary exhibition, Ethnological Museum Berlin.
(Photo © Pascal Querner)

the Museum of European Cultures were known to have an active insect infestation (Figs 2–6); other collections of the foundation were monitored as part of the introduction of an IPM concept. Only a few of the Foundation's collections already had insect monitoring (some parts of the Ethnological Museum, the textile collection of the Museum of European Cultures and the Bode Museum). Monitoring started in spring 2010 (February/March) using sticky blunder traps and pheromone traps (mainly for clothes moth). Traps were placed primarily in storage areas. Table 1 lists the museums sampled and shows the type and number of traps used in each. All traps were checked regularly by the IPM appointee together with the conservators of the collections and the insect pests identified. Pest species were kept for the reference collection of the Foundation.

Results

A total of 14 insect pest species were found in all the museums investigated (Tables 2 and 3). The most common and abundant pest species were webbing clothes moth, *Tineola bisselliella*, varied carpet beetle, *Anthrenus verbasci*, brown carpet beetle, *Attagenus smirnov,* and the khapra beetle (also called Berlin beetle), *Trogoderma angustorum.*

Pests were found in almost all the collections during the one-year monitoring programme (*see* Table 2); but not all of them indicated an active infestation of objects, some are as a result of poor housekeeping and a lack of cleaning (for example in the Bode Museum, Photographic Museum and Kupferstichkabinett). Here, actions were taken to improve cleaning of the floors and window-sills. Some pest species were found in collections where an infestation was not expected, such as silverfish, *Lepisma saccharinam,* in the Geheimes Staatsarchiv, biscuit beetle, *Stegobium paniceum,* in the Kunstbibliothek (Fig 7), clothes moth, *Tineola bisselliella,* in the Museum of Musical Instruments (Fig 8), and the brown-dotted clothes moth, *Niditinea fuscella,* in the Museum of Islamic Art.

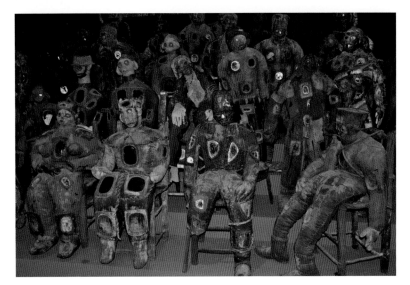

In the Ethnological Museum a new species of pest was found that was not known before in this collection, *Anthrenus pimpinellae* (Fabricius 1775); but the expected high infestation of tobacco beetle, *Lasioderma serricorne,* was not found. The areas with the greatest infestation problems are in the Ethnological Museum, with biscuit beetle in the African and American collections, with clothes moth and carpet beetle in the Musical Instrument collection, and with pests generally in the Museum of European Cultures.

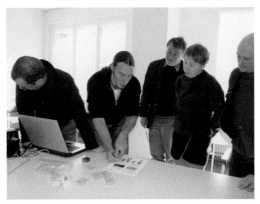

Fig 5 (left)
Training museum staff with insect pest identification courses at the Ethnological Museum Berlin.
(Photo © Pascal Querner)

Fig 6 (below)
Infested wood objects from the Museum of European Cultures Berlin.
(Photo © Pascal Querner)

Discussion

It can be shown that the Foundation made the right decision to appoint a 12-month IPM position instead of implementing a sulfuryl difluoride fumigation. A larger number of museums could be investigated, the most abundant pest species and problems were identified, and solutions were discussed with the heads of the museums and the conservators. At the end of the 12 months, presentations were given in all the collections to discuss further actions. To implement a full IPM programme in 12 months is difficult, as results from treatments and changes can often only be seen over several years. In the Ethnological

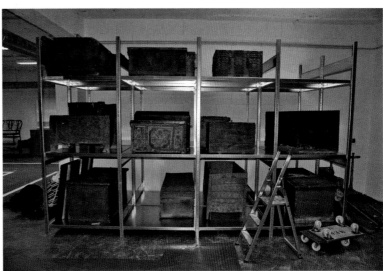

Table I Museums of the Foundation investigated, type and number of traps used in the collections

Museums	Sticky blunder traps	Pheromone traps for clothes moth	Pheromone traps for biscuit beetle
Ethnological Museum (EM)	312	85	29
Museum of European Cultures (MEC)	150	51	–
Bode Museum (BM)	96	23	–
Museum of Islamic Art (IM)	68	26	–
Egyptian Museum (EM)	53	9	–
Kunstbibliothek (KB)	94	–	–
Museum of Photography (MP)	15	–	–
Kupferstichkabinett (KK)	40	2	–
Museum of Musical Instruments (MI)	36	8	–
Geheimes Staatsarchiv (GS)	56	–	–

Table 2 Insect pest species found in different museums of the Foundation

Detected pest species	EM	MEC	BM	EM	IM	PM	MI	KK	KB	GS
T. bisselliella	114	74	48	2	5	–	20	I	–	–
N. fuscella	–	–	–	–	8	–	–	–	–	–
A. verbasci	66*	33	17	9	12	3	1	5	10	23
Attagenus sp.	13	2	–	10		–	–	–	3	1
D. maculates	–	2	–	–	–	–	–	–	–	–
T. angustum	14	45	–	1	1	–	–	–	–	–
S. paniceum	84	2	–	–	–	–	–	–	I	–
A. punctatum	15	5	–	–	–	–	–	–	–	–
P. tectus	5	2	I	–	I	–	–	2	–	–
L. saccharina	–	†	†	–	†	–	†	–	†	†

* Besides *A. verbasci*, *A. scrophulariae* and *A. pimpinellae* were also found.
† A high presence of silverfish in the collection.

Table 3 Insect pest species found in the collections of the Ethnological Museum

	S. paniceum	Anthrenus sp.	T. bisselliella	A. punctatum	Attagenus sp.	T. angustum	Ptinus t.
Africa	23	12	2	–	–	–	–
American Ethnology	61	20	11	–	1	–	–
East Asia	–	4	31	2	–	–	–
American Archaeology/ Bootshalle	–	11	13	13	2	6	–
American Ethnology (mummies)	–	1	1	–	–	–	1
Cellar	–	–	1	–	–	–	–
Pacific	–	1	10	–	1	–	1
South East Asia	–	2	*	–	–	–	–
Islam	–	–	6	–	–	*	–
N2 Treatment room	–	–	+	–	–		–
Music instruments	–	15	21	–	7	8	3
Vodou exhibition	–	–	48	–	2	–	–

* An active infestation was found in the collection but no pests on the monitoring traps. *Anthrenus* sp. are mainly *Anthrenus verbasci*, but also some individuals of *Anthrenus scrophulariae* and *Anthrenus pimpinellae*. *Attagenus* sp. are *Attagenus smirnovi* and *A. pellio*. In recent months *Nicobium castaneum* (Olivier 1790) has also been found at the Ethnology Museum site.

Museum, problems concerning the building, climate control, lack of a quarantine policy, the problematic functioning of the nitrogen tent, and the lack of personnel to prepare and pack objects for the freezing chamber, were presented to the head of the Ethnological Museum. The acquisition of a larger nitrogen chamber was suggested as a way to tackle the large number of infested objects within their own collections, and to help cope with the growing demand for pest treatments within the rest of the museums in the Foundation.

Only a few museums or collections in the world can afford a full-time IPM position, and often the work of IPM is dealt with by the conservators alongside their many other tasks. English Heritage (UK), with currently 62 sites where an IPM programme is in place, is probably one of the few other large collections besides the Foundation in Berlin to fund a full-time IPM position (Lauder nd). As to the question: 'what size does a museum or collection have to be to require a full-time or part-time IPM expert post?' the answer depends on the problems in place and the other responsibilities of the individual. In Berlin, besides implementing IPM in nine museums, a large number of workshops were held to inform and train museum staff, as well as talks presented to the public. In addition, a research proposal was written and a questionnaire on IPM prepared. The IPM consultant of the Foundation also helped to train staff and implement an IPM programme at the Georgian National Museums as part of a European Union (EU) Twinning project (Querner *et al* 2011).

Conclusion

For large collections such as in the National Museums of Berlin or the Prussian Cultural Heritage Foundation, a full-time IPM position is needed to cope with the demand of the large number of different museums. In the first year, nine museums were included in the IPM programme and time was made for research, writing publications, public relations, talks and workshops. Having an IPM expert within the institution was well received by the conservators in the Foundation. Smaller museums may only need to appoint a part-time position or, to save money, contract an external IPM expert or consultant.

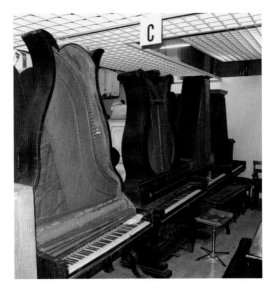

Fig 7 (left)
Storage site of the Kunstbibliothek (Art Library) in Berlin where one individual of biscuit beetle was found. (Photo © Pascal Querner)

Fig 8 (below)
Storage depot of the Musical Instrument Museum, where clothes moth were found. (Photo © Pascal Querner)

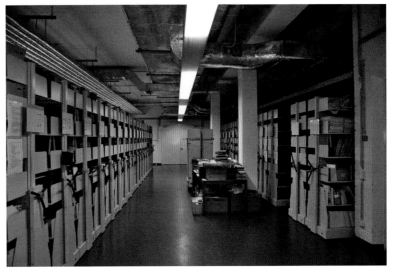

Acknowledgements

We thank the Prussian Cultural Heritage Foundation for supporting an IPM position for one year and all the colleagues in the museums for their cooperation.

References

Ackery, P R, Pinniger, D B and Chambers, J 1999 'Enhanced pest capture rates using pheromone-baited sticky traps in museum stores'. *Studies in Conservation* **44**(1), 67–71

Åkerlund, M, Bergh, J E, Stenmark, A and Wallenborg, I (eds) 1998 *Proceedings of the 3rd Nordic Symposium on Insect Pest Control in Museums, 24–25 September 1998.*

Stockholm: PRE-MAL and ICOM Swedish National Committee

Albert, G D and Albert, L M 1988 'Integrated pest management: A program for museum environments', *in* Zycherman, L A and Schrock, J R (eds) *A Guide to Museum Pest Control.* Washington DC: The Foundation of the American Institute for Conservation of Historic and Artistic Works and Association of Systematics Collections, 169–73

Boylan, P J (ed) 2004 *Running a Museum: A Practical Handbook,* Paris: ICOM International Council of Museums

Brand, S J and Wudtke, A 1997 'Bekämpfung von Textilschädlingen mit Kohlenstoffdioxid'. *Restauro, Fachzeitschrift für Kunsttechniken, Restaurierung und Museumsfragen* 4, 272–6

Brokerhof, A, van Zanen, W B and den Teuling, A J M 2007a *Fluffy Stuff: Integrated Control of Mould in Archives.* Amsterdam: Netherlands Institute for Cultural Heritage (ICN) and IADA

Brokerhof, A, van Zanen, W B, van de Watering, K. and Porck, H 2007b *Buggy Biz: Integrated Pest Management in Collections.* Amsterdam: Netherlands Institute for Cultural Heritage (ICN)

Child, R E and Pinniger, D B 1993 'Insect trapping in museums and historic houses', *in* Wildey, K B and Robinson, W H (eds) *Proceedings of the 1st International Conference on Insect Pests in the Urban Environment,* Cambridge: ICIPUE, 267–70

Cox, P D, Pinniger, D B and Mueller, D 1996 'Monitoring populations of the webbing clothes moth, *Tineola bisselliella,* using pheromone lures', *in* Wildey, K B and Robinson, W H (eds) *Proceedings of the 2nd International Conference on Insect Pests in the Urban Environment,* Edinburgh: ICIPUE, 541–5

Gilberg, M 1989 'Inert atmosphere fumigation of museum objects'. *Studies in Conservation* 34(2), 80–84

Gilberg, M 1991 'The effects of low oxygen atmospheres on museum pests'. *Studies in Conservation* 36(2), 93–8

Gilberg, M and Roach, A 1991 'The use of a commercial pheromone trap for monitoring *Lasioderma serricorne* (F.) infestations in museum collections'. *Studies in Conservation* 36(4), 243–7

Jessup, W C 1998 'Integrated pest management into operation'. *Collections Caretaker* 1(3), 1–8

Kingsley, H, Pinniger, D, Xavier-Rowe, A and Winsor, P 2001 *Integrated Pest Management for Collections, Proceedings of 2001: A Pest Odyssey.* London: James and James

Lauder, D nd The Integrated Pest Management Programme (Online poster). London: English Heritage. www.english-heritage.org.uk/content/imported-docs/f-j/ipm_poster (accessed 22 June 2011)

Linnie, M J 1987 'Pest control: A survey of natural history museums in Great Britain and Ireland'. *The International Journal of Museum Management and Curatorship* 6, 277–90

Linnie, M J 1996 'Integrated pest management: A proposed strategy for natural history museums'. *The International Journal of Museum Management and Curatorship* 15(2), 133–43

Maekawa, S. and Elert, K 2003 *The Use of Oxygen-free Environments in the Control of Museum Insect Pests.* Los Angeles: The Getty Conservation Institute

Noldt, U and Michels, H 2007 *Holzschädlinge im Fokus – Alternative Maßnahmen zur Erhaltung historischer Gebäude.* Beiträge der internationalen Tagung im LWL-Freilichtmuseum Detmold/Westfälisches Landesmuseum für Volkskunde, 28–30 der Juni 2006, Detmold, Merkur Verlag.

Pinniger, D B 1998 'Integrated pest management (IPM) in museums', *in* Repp, B, Stäbler, W and Wießmann, A (eds) *Das Museumsdepot: Grundlagen – Erfahrungen – Beispiele.* MuseumBausteine Band 4. Weltkunstverlag München, 111–23

Pinniger, D B 2003 'Saving our treasures – controlling museum pests with temperature extremes'. *Pesticide Outlook* 14(1), 10–11

Pinniger, D B 2004 *Pest Management in Museums, Archives and Historic Houses.* London: Archetype Publications Ltd

Pinniger, D B 2008 *Pest Management: A Practical Guide.* Cambridge: Collections Trust

Pinniger, D B, Child, R E and Chambers, J 2004 'Attractant pheromones of museum insect pests'. *Australian Institute for the Conservation of Cultural Material Bulletin* 28, 4–10

Pinniger, D B and Winsor, P 2004 *Integrated Pest Management: A Guide for Museums, Libraries and Archives.* London: Museums Libraries Archives Council

Querner, P and Morelli, M 2010a 'Integrierte

Schädlingsbekämpfung in Museen – Erfahrungen einer Umstellung'. *Restauro, Fachzeitschrift für Kunsttechniken, Restaurierung und Museumsfragen* **4**, 234–41

Querner, P and Morelli, M 2010b 'Leitfaden für eine Einführung und Umstellung zur Integrierten Schädlingsbekämpfung (IPM)'. *Restauro, Fachzeitschrift für Kunsttechniken, Restaurierung und Museumsfragen* **5**, 332–3

Querner, P, Kalandadze, N, Khoshtaria, V, Phakadze, V, Nawroth, M and Simon, S 2011 Developing and implementing an Integrated Pest Management concept in the Georgian National Museum, in cooperation with the National Museums in Berlin. Poster presentation at *Pest Odyssey 2011: Ten Years Later Conferences, London, 26–28 October 2011*. London: English Heritage

Ranacher, M 1998 (ed) 'Mikroorganismen und Schadinsektenbefall im Depot: Ursachen, Sanierung, Hygiene und Gesundheitsschutz', *in Das Depot – der andere Teil der Sammlung*. Landesstelle für die Nichtstaatlichen Museen, **9**. Bayerischer Museumstag, München, 53–74

Reichmuth, C, Unger, W and Unger, A 1991 'Stickstoff zur Bekämpfung Holzzerstörender Insekten in Kunstwerken'. *Restauro, Fachzeitschrift für Kunsttechniken, Restaurierung und Museumsfragen* **4**, 246–51

Reichmuth, C, Unger, W and Unger, A, 1994 'Bekämpfungsmaßnahmen mit Stickstoff oder Kohlenstoffdioxid'. *Der Praktischer Schädlingsbekämpfer* **46**, 81–7

Story, KO 1986 *Approaches to Pest Management in Museums*. Washington DC: Conservation Analytical Laboratory, Smithsonian Institution, 85–101

Strang, T J K 1992 'A review of published temperatures for the control of pest insects in museums'. *Collection Forum* **8** (2), 41–67

Strang, T J K 1999 'A healthy dose of the past: A future direction in herbarium pest control?' *in* Metsger, D A and Byers, S C (eds) *Managing the Modern Herbarium: An Interdisciplinary Approach*. Society for the Preservation of Natural History Collections. Vancouver: Elton-Wolf Publishing, 59–80

Strang, T J K and Kigawa, R 2006 'Levels of IPM control: Matching conditions to performance and effort'. *Collection Forum* **21**(1–2), 96–116

Unger, W, 2002 'IV Schädlingsbekämpfung', in Hilbert, G S (ed) *Sammlungsgut in Sicherheit*. Berlin: Gebrüder Mann Verlag, 290–332

Unger, W, Fritsche, H and Unger, A 1996 'Zur Resistenz von Materialien und Stabilisierungsmitteln für Kunst- und Kulturgut gegenüber Holzzerstörenden Insekten'. *Zeitschrift für Kunsttechnologie und Konservierung* **10**(1), 106–16

Wießmann, A 1998 'Schädlinge in Museen und ihre Bekämpfung', *in* Repp, B, Stäbler, W and Wießmann, A (eds) *Das Museumsdepot: Grundlagen – Erfahrungen – Beispiele*. MuseumBausteine Band 4. Weltkunstverlag München, 105–09

Wudtke, A 1996 'Integrierte Schädlingsbekämpfung in Museen. Entomologentagung in Göttingen 1995'. *Mitteilungen der Deutschen Gesellschaft für Allgemeine und Angewandte Entomologie* **10**, 251–3

Wudtke, A 2002 *Möglichkeiten des Methodentransfers vom Vorratsschutz zum Materialschutz: Bekämpfung von Museumsschädlingen am Beispiel der Kleidermotte Tineola bisselliella (Hum. 1823), Leipidoptera: Tineidae*, Auflage: Mensch & Buch (Dissertation, Humboldt-Universität Berlin)

Wudtke, A 2003 'Museumsschädlinge – Vermeidung und Bekämpfung am Beispiel der Kleidermotte'. *Museumskunde* **68**(2), 122–8

7

Building with pest management in mind: a case study from the Canadian Museum of Nature

Marcie Kwindt* and Laura Smyk

*Canadian Museum of Nature, Ottawa, Ontario, Canada,
LSmyk@mus-nature.ca (author for correspondence)
http://nature.ca/en/home

ABSTRACT

In 1997 the Canadian Museum of Nature (CMN) moved its collections, research labs and administrative offices to one central location, the Natural Heritage Building (NHB) in Gatineau, Quebec. Efforts were made to include preventive conservation principles into the actual design of this building. Upon completion, the NHB was deemed state of the art, having many integrated pest management (IPM) related building features incorporated into its design. Now, 14 years later, these IPM features are re-evaluated and information is shared about which features are effective and what can be done to make less effective features work better. Information on IPM features that have been added or changed since the opening of the NHB is also included.

KEYWORDS

Building design, building envelope, integrated pest management, preventive conservation

Introduction

The Canadian Museum of Nature (CMN) currently calls two main buildings home: the Natural Heritage Building (NHB) and the Victoria Memorial Museum Building (VMMB), where visitors can view specimens on display. Prior to the 1997 opening of the NHB, the CMN staff and collections were spread out over more than 14 buildings throughout the Ottawa capital region. The much anticipated move to one central location was a chance for the CMN to incorporate principles of preventive conservation into the actual building design with the end result being the NHB, an exemplary model of how to plan a building to house museum collections.

The NHB was designed with many integrated pest management (IPM) related building features (Canadian Museum of Nature. 1996a,b,c). Many of these resulted from consultations with CMN collections and conservation staff, as well as advice from outside sources such as the Canadian Conservation Institute (CCI). Over the past 14 years many of these features have continued to work as originally intended, with some requiring adjustments in order to gain their optimum functionality. Additional IPM features that were not originally considered at the time the NHB was being built have also been added (Canadian Museum of Nature Conservation Section 2003, 2006; Kwindt nd; Waller 2000).

Exterior of the NHB

The building envelope

A building envelope can be defined as that which separates the exterior environment and the interior environment of a building. The physical components of a building envelope consist of the foundation, walls, roof, windows and doors. The building envelope of the NHB was created to be the primary protective barrier between pests and the valuable national collections it holds inside. A comprehensive list of the IPM-related building design features that were added or changed from the original design can be found in Table 1.

- *Foundation/Walls/Roof.* Efforts were made to ensure that all cracks and holes in the exterior of the building were properly sealed. This included sealing around all conduit lines, piping and ventilation ducts coming into, and going out of, the NHB. In addition, all vents were fitted with screens to prevent pests from gaining access to the building. If pests do make it past the first set of screens they are eventually trapped in the second level of screening found within the various mechanical rooms.

- *Exterior doors.* All main entrances to the NHB utilise two sets of doors. All doors to the building are equipped with tight-fitting sweeps on the bottom and weather stripping along both sides and across the top to prevent any gaps that pests could gain access through. These IPM-related features are successful although they do require regular inspection and replacement as normal wear and tear will reduce their effectiveness.

- *Loading dock cargo bay doors.* A few years after moving into the NHB, nesting and roosting activity by pigeons on top of the cargo bay door was discovered. Bird netting (wire screening) was installed above the doors. This was meant to discourage bird activity and decrease the chance of other pests that are routinely found in nesting sites, such as dermestids and clothes moths. It was a successful addition, as it rectified the problem; however, regular inspection and maintenance are necessary to ensure its continued functionality.

- *Windows.* The NHB was designed with windows that do not open. As far as IPM-related design is concerned this is the best possible scenario for reducing the introduction of pests to the building via windows. It also helps control the internal environment of the NHB.

- *Lighting.* In most cases, an effort was made to position lights away from entrances to the building and to also direct light away from the entrances. A move was also made to using sodium vapour lighting which is less attractive to pests and saves energy. With CMN attempts to save energy and money, a decrease in the amount of interior lights visible at night from the exterior of the NHB was made. This also doubles as an IPM building feature because fewer pests will be attracted to the building as a result.

Landscaping and the site around the building envelope

An integral part of the NHB design and IPM deals with the landscaping and maintenance of the environment in close proximity to the building.

- *Vegetation-free zone.* A one metre wide vegetation-free zone surrounds the entire perimeter of the building exterior. It is constructed from durable landscape cloth covered by pea gravel with a metal barrier running along its border. This vegetation-free zone allows for easier visual inspections whilst promoting better air circulation along the building walls. Additional IPM advantages are that it prevents rodent burrows, plant colonisation and insect harbourage. To ensure its effectiveness regular maintenance in the form of weeding is required.

- *No window boxes or elevated planters.* Although they can be attractive in appearance, window boxes and elevated planters were left out of the final NHB design due to their ability to provide excellent harbourage sites for pests such as rodents. Elevated planters also have a tendency to accumulate debris and food trash which in turn would attract additional pests.

- *No building contact by trees or shrubs.* Vegetation that comes in contact with any part of the building can be used by climbing pests to gain access to it. Trees and shrubs this close to the building also interfere with visual inspections whilst increasing moisture content along the foundation walls and providing shelter for pests. Regular pruning is necessary to keep this IPM building design feature working effectively.

Table 1 IPM-related building design features at the Natural Heritage Building showing their incorporation, function, effectiveness and improvement

IPM related building features	Part of Original Design	Function of IPM building design feature	Why an addition or change to the original design was necessary	IPM feature functioning effectively	How to improve the effectiveness of IPM building design feature
The NHB Exterior					
Doors:					
properly fitted door sweeps	Y	keeps pests out	Changing the door sweeps is part of regular building maintenance.	Y	
loading dock cargo door equipped with bird netting	N	pest deterrent while cargo doors are open	A few years after opening we had some problems with birds nesting in the space atop the loading dock doors. The netting was added and we have been incident free since then.	Y	
Landscaping:					
vegetation free zone	Y	pest deterrent; visual inspection aid; protects foundation from roots; better air circulation along walls	Over time seeds and soil fall into the vegetation free zone. If not weeded on a regular basis this zone can become full of plant life and loose its effectiveness as an important IPM building feature. This is what has been happening to our vegetation free zone over time.	N	maintenance (weeding) of this area is necessary for it to work effectively; appropriate funds need to be allocated for upkeep
no tree or shrub should come in contact with any part of building	Y	vegetation shelters pests; easy access for climbing pests; fire risk; difficult visual inspections; increases moisture content along foundation walls	The trees and shrubs have been allowed to naturalize. Although they were an appropriate distance away from the building when first planted they are now coming into contact with the building.	N	maintenance (pruning) of the trees and shrubs on the grounds are necessary; appropriate funds need to be allocated for upkeep
Lights:					
sodium vapour bulbs used	N	lighting is less attractive to insects	A change was made to sodium vapour lighting primarily for their energy efficiency. An added bonus has been they are less attractive to pests.	Y	
reduce interior lighting visible to the outside	N	helps reduce amount of pests attracted to building	With the need to reduce energy consumption came the turning off of many lights that had once been visible during the night hours. This energy saving reduction also had the effect of reducing the number of pests being attracted to our building.	Y	
Interior of NHB					
Overall Building Layout:					
office spaces away from collections	Y	keeps pests out; areas with highest chance of pests kept farthest away from collections	For a variety of reasons many staff members who also have space available in lab areas have taken to working primarily in these locations. As these lab spaces begin to function more as office spaces IPM procedures are often forgotten.	N	A space allocation project is currently underway to better utilize the office and lab spaces available. This should eliminate the need for some staff to treat lab space as their primary office.

Feature		Function	Comments		Recommendation
unpacking room	Y	keeps pests out; corrugated cardboard and packing materials stored here.	Was intended to be a well organized area with users responsibly discarding dirty packing material and storing reusable materials neatly. Multiple users and a lack of responsibility among them for this space have left it a constant mess.	N	Increase in staff awareness of IPM should help. We are working towards the implementation of required yearly IPM orientation and training specific to this room for all users.
temporary holding room	X	keeps pests out; unclean items waiting to be dealt with stay here	Originally designed to be a "temporary holding area" while items wait for available freezer space, etc. This room has become a dumping ground and is now full of field gear and miscellaneous items.	N	An annual cleaning of this space with unclaimed items being discarded/recycled should be implemented
CO2 bubble room	Y	keeps pests out; was to be used for IPM purposes	This space was never used effectively for its intended purpose. It has recently been turned into a "dirty lab" space allowing researchers to work with material not able to enter the collection and lab corridor because it hasn't passed through the proper IPM procedures	N	Although not being used effectively for its originally intended purpose this room's designation as a "dirty lab" is the positive result of having an effective IPM program in place.
Doors:					
sweeps	Y	keeps pests out	Door sweeps were removed and replaced by automatic door bottoms (ADB). The idea was great but the doors now no longer close properly. This results in some of them being left ajar and creating a greater opening for pests to enter than if the original sweeps had been left on.	N	Although costly, the current sweeps should be replaced. If not now, when replacements are needed properly fitted ones should be used. Better communication between museum departments would have prevented this issue before it became a problem.
double doors	Y	keeps pests out	Door thresholds were removed to allow for smoother transport of carts in and out of collection and lab rooms.		Ensuring all doors are equipped with properly working door sweeps will effectively block any pests.
Cabinets:					
good seals on cabinet doors	Y	keeps pests out; can also work to keep pests in	Many cabinets with poor seals made the move to the NHB	Y	Time and budgets permitting, old seals are removed from cabinets and replaced with new ones.
new cabinets upon move	Y	keeps pests out; can also keep pests in; clean and fitted with good gaskets	Due to budgetary restraints new cabinets could not be purchased for all collections	Y	Time and budgets permitting, new cabinets have continued to be added to collection pods as needed.
Shelving:					
open and visible	Y	aids in visual inspection; no pest harbourage or dark areas for pest breeding; removes specimens form floor areas		Y	Time and budgets permitting, open shelving continues to be used and is added to collection pods when needed.
Miscellaneous:					
zap traps installed in office areas	N	attracts wasps and other flying pests; eliminates any that enter trap	these were deemed necessary to help deal with wasps in office areas.	N	It is crucial that the cleaning of these traps be added to existing housekeeping procedures to avoid the dead insects attracting additional pests.

* For a comprehensive listing of all IPM related building design features contact Laura Smyk at the Canadian Museum of Nature.

• *Ground slopes away from building.* The design of the NHB called for the ground to slope away from the building exterior upon completion. This gradual slope away from the exterior walls eliminates the possibility for areas of standing water to be created. This results in no areas of high moisture forming along the building foundation which in turn decreases chances of mould and mildew growth that have been known to support pests. Over time the ground can settle and the slope can begin to erode. Annual inspections and maintenance, to ensure that the ground continues to slope away from the building, help to confirm that this remains a useful IPM-related building feature.

• *Well-maintained site free of clutter and garbage.* Outdoor maintenance is crucial to ensuring a building's IPM programme is working efficiently. The CMN does not allow the use of pesticides on any of its sites. By keeping the NHB site weeded, lawns cut, plantings pruned and the area free of clutter and garbage, great strides are being taken to discourage pests and contribute to visual inspections. However, some of the necessary pruning and weeding have not been done in the last few years. The conservation section continues to encourage this important work to be reinstated before pest-related problems arise.

Interior of the NHB

NHB layout

The NHB was designed to strategically separate certain areas of the building and incorporate particular features to assist with pest management. A floor plan highlighting the main floor layout of the NHB and key IPM features can be found in Figure 1.

• *Three separate collection pods.* The collections are all stored in the same area of the NHB. Each collection pod is surrounded by corridors and further divided into smaller collection rooms within each pod. No collection pod has exterior windows or walls, which significantly reduces the risk of pests.

• *Dynamic buffer zone (DBZ) around collection pods.* Between the exterior walls of the NHB and the collection pod walls there is a dynamic buffer zone. This is essentially a narrow corridor that helps with environmental control and pest management. If pests do gain access to the building they have to cross the DBZ before they reach interior doors leading to the collection pods. These interior doors are for use as emergency exits only. This feature has shown to be very successful. We monitor both sides of the interior doors (DBZ and pods) and there is a significant difference in the number of pests discovered in the DBZ compared to the number found in collection pods, with the DBZ accounting for the largest number.

• *Single storey collection storage and research labs.* No second storey above the collection pods and research labs reduces the risk of pests entering via connections to above floors.

• *Cafeteria away from collections.* The cafeteria was designed to be completely separate from all staff working areas and the collection pods. The cafeteria being located in its own wing provides a structural barrier via walls, as well as one of distance, between it and the collections. Although staff are permitted to eat at their desks, they are provided with sealed plastic containers to house food, are encouraged to store lunches in available fridges and to eat in the cafeteria.

• *Offices and research labs away from collections.* The offices for collections and research staff are in a separate area away from the collection pods and research labs. Not having the offices and labs incorporated into the collection storage areas decreases the chance of pests being inadvertently brought in by staff in bags or on shoes. This was part of the original design; however, for a variety of reasons many staff members who also have space available in lab areas have taken to working primarily in these locations. As these lab spaces begin to function more as office spaces IPM procedures are sometimes being forgotten. A space allocation project is currently underway that aims at better utilising the office and lab spaces available to staff. This should eliminate the need of some staff to treat lab space as their primary office.

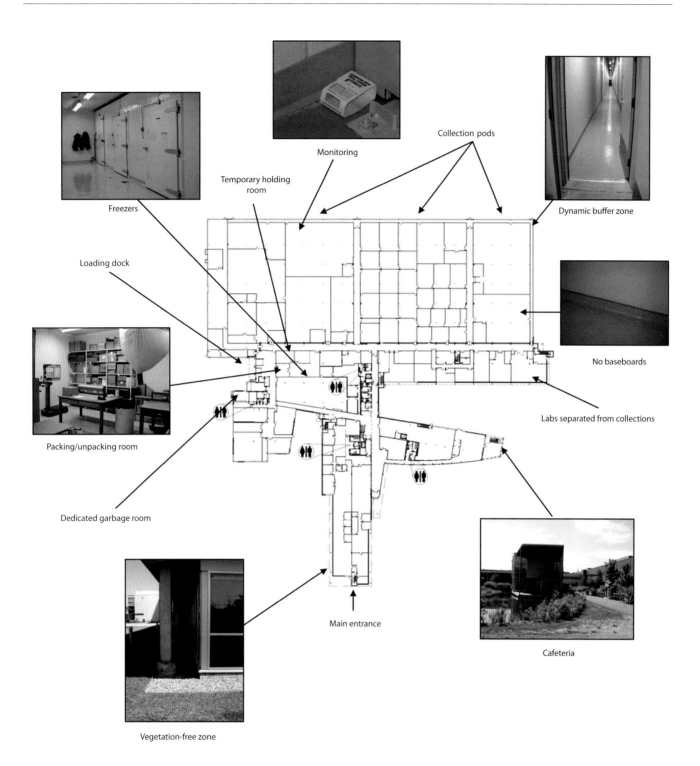

Monitoring

Collection pods

Temporary holding room

Freezers

Dynamic buffer zone

Loading dock

No baseboards

Labs separated from collections

Packing/unpacking room

Dedicated garbage room

Main entrance

Cafeteria

Vegetation-free zone

Fig 1
Floor plan of the Natural Heritage Building highlighting IPM features. (Images © Canadian Museum of Nature)

- *Central security desk.* All visitors and staff must pass by security upon entering the NHB. They are trained to provide all visitors with a brief orientation that includes not bringing bags, food or drink into the collection areas.
- *Shipping and receiving separated from collection pods.* All deliveries, from specimen loans to building supplies, are delivered to the NHB via the loading dock. They are then delivered to the proper destination and processed according to IPM procedures if necessary.
- *Dedicated garbage room off loading dock.* All garbage that is collected in the building is stored in an isolated, sealed room off the loading dock and away from collection areas until pick-up.

Dedicated IPM areas

These consist of:

- *Packing/unpacking room.* This room is directly off the loading dock. All specimens travelling into the building move through this room where they are unpacked from boxes and crates. It is designed to prevent possibly infested packing material from moving into the collection areas. Any specimens that are destined for the collection pods or labs that cannot be frozen are inspected in this area. Although a great feature, it tends to become a habitat for pests due to packing material and boxes piling up and making the area difficult to clean. Sticky trap inspection consistently shows pest activity in this room.
- *Freezers.* Directly off the packing/unpacking room is a room that contains three walk-in freezers. As a precautionary measure all incoming materials destined for the collection pods and labs are cold treated for seven days at $-30°C$. One freezer has proven sufficient for such-short term treatments. The other two freezers now hold Collection and Research material for longer periods of time. Materials not suitable for freezing are visually inspected and may undergo an alternative IPM procedure.
- *CO_2 bubble.* When the NHB was built a CO_2 bubble was purchased and a separate room was designed to house it. Unfortunately, the bubble was a large, rigid frame model that proved to be oversized relative to our ongoing requirements. Due to its operating costs and low request levels a new home was eventually found for it. On the few occasions when it has been necessary to treat specimens that cannot undergo freezing, the CO_2 bubble at the Canadian Museum of Civilization has been used with good results. This room is now being used as a dirty lab for researchers to look at materials brought in from field work that cannot be put through the freezers or allowed to enter into the collections and labs section of the NHB. This has proven to be a valuable addition to our IPM program.
- *Temporary holding room.* This area is designed to house specimens that cannot be treated or inspected in a timely fashion, and therefore cannot be brought into the collection areas. In theory, this is a great feature. However, the reality is that this room has a tendency to become a dumping ground for all sorts of materials, not just specimens, and not so temporary. This feature requires a high level of organisational discipline to remain useful for its intended purpose.

Collection pods

Within the collection pods, specific IPM-related building design features assist in preventing, monitoring, and controlling pests.

- *Four separate HVAC zones.* The collections are divided into four separate heating, ventilation and air conditioning (HVAC) zones based on environmental requirements. These separate HVAC zones result in no mixing of air between the different zones and no chance of pests migrating through the ductwork from one zone to another. In addition, these four zones are all completely separate from the rest of the NHB. The different HVAC zones are used to help us better develop our IPM monitoring programmes.
- *Designated fur vault.* The NHB was designed with a fur vault in order to house what is one of the most susceptible collections for pest infestation. The fur vault is kept at $10°C$ to discourage pests

from entering the space and to slow them down if they do.

- *Restricted access.* All individuals allowed access to the collections and labs areas must first receive an orientation that includes pest management policies and procedures.

Interior doors

- *Sweeps.* All collection pod doors had high quality sweeps to prevent pests from gaining access through any gaps underneath. Originally all doors had thresholds but pushing carts over these caused an increase risk of damage to specimens and they were therefore removed. To eliminate the gap created, the original sweeps were replaced with automatic door bottoms (ADB). The idea was a good one but due to incorrect installation many of the doors no longer close properly. This has resulted in some of them being left ajar creating a greater opening for pests to enter than if the original sweeps had been left on and it is now recommended that the sweeps are re-fitted.

- *Double doors to collection pod.* Double doors leading to the collection pods and labs help prevent any pests that make it into this section of the NHB from migrating in from the corridors.

Interior walls, ceiling and floors

Proper wall and ceiling choices assist in pest management.

- *Light colour on walls.* The decision to have all walls decorated in a light palette was primarily to assist with visual inspections.

- *No suspended ceilings.* The open ceilings and exposed pipe/duct work eliminates areas that are hard to clean, prevents build up of flies which could be a food source and eliminates a possible breeding area. This feature also aids in visual inspection.

- *Zap traps.* Zap traps were placed on the walls by the offices of the collection and research staff to help control flies and wasps. The traps are a concern because if they are working effectively they accumulate dead insects which can then attract additional insects such as dermestids. It is necessary to have a regular cleaning-out of these zap traps added to the housekeeping regime for the NHB.

- *Light coloured flooring.* The light colour and low pattern flooring in the NHB helps with spotting pests. All collection pods, labs, corridors and collection and research offices incorporate this feature.

- *No carpeting.* Floors covered in carpeting hinder the detection of pests, gather possible food sources and make cleaning spills more difficult. The only area in the whole NHB where carpet can be found is in the administrative work areas. There was a managerial decision made during the design phase that the risk to the collections from synthetic carpeting in the administrative areas is a manageable risk. This carpeting is low-pile.

- *No baseboards in collection pods or labs.* All collection pods and labs have floors that curve up along the wall with no baseboards (skirting). This reduces cracks and crevices for pests to inhabit and helps facilitate proper cleaning.

Cabinets and shelving

- *New cabinets.* When the Birds and Mammals collections were moved into the NHB the old cabinets were left behind. Everything was moved into new, clean cabinets. This was to prevent the possibility of pests moving in with the old cabinets. When needed, and as budgets allow, the NHB continues to work toward re-housing its other collections in new, pest-free cabinets.

- *Good seals on cabinets.* Well-sealed cabinets prevent pests from gaining access to cabinets and, if there is an infestation, from escaping and going elsewhere. The new cabinets obtained for the move to the NHB came with new gaskets. Seals on older cabinets are being replaced with new ones as time and money allow.

- *Cabinets moveable by forklift.* The cabinets in the collections that are deemed the most susceptible to pest infestation (Botany, Birds and Mammals) are moveable by forklift and fit into our walk-in freezers. If an infestation is ever discovered the entire cabinet can be easily treated. This feature has only been utilised once when a cabinet from the Bird collection was placed in the freezer. The operation did work as planned.

- *Open shelving*. Open shelving in collection rooms is beneficial for visual inspections. The open shelving discourages dark areas popular for pest breeding and reduces any harbourage of pests.

Conclusion

Since its inception, the CMN staff involved with the design of the NHB strove to incorporate preventive conservation measures, in the form of IPM-related features, into the building plan. Fourteen years later, it could be argued that the NHB continues to remain a model for how to design a building to house museum collections. Some changes and additions to the original building design have taken place over the years, but the NHB remains a vast improvement over the various buildings that were once home to the collections. The IPM building features need to be reviewed and reconsidered on a regular basis and building maintenance and inspections must continue, with changes being made when necessary.

Based on first-hand knowledge of the NHB, and involvement with facilitating the CMN's IPM program, the most serious pest management problems encountered at the NHB have been due to human elements rather than the result of building deficiencies. However, the discovery that people are the biggest hindrance to running an efficient IPM programme at the NHB does not equate to a reduction in upkeep necessary to maintain all IPM-related building features.

Experience has shown that the best weapon of defence is knowledge, which highlights the need to continue to deliver IPM orientation and training, attempt to enforce IPM procedures already in place and better communicate with staff the importance of having a strong IPM program. Staff participation is the key to a successful IPM programme and by disseminating IPM knowledge to individuals using the NHB, it can empower them to help run an even more effective IPM program.

Acknowledgements

The authors would like to thank Martin Leclerc of the Canadian Museum of Nature and Dr Robert Waller of Protect Heritage Corporation for their contributions and guidance. We are grateful to the Canadian Conservation Institute and especially to Tom Strang for expert guidance in developing specifications during the design of the NHB.

References

Canadian Museum of Nature 1996a 'Notes from CMN Pest Management Committee, June 11, 1996'. Unpublished internal report

Canadian Museum of Nature 1996b 'Integrated Pest Management Committee. Minutes of meeting of October 8, 1996'. Unpublished

Canadian Museum of Nature 1996c 'Mandate of the Pest Management Committee (draft)'. Unpublished internal report

Canadian Museum of Nature Conservation Section 2003 'Canadian Museum of Nature: 2003 Natural Heritage building risk assessment update'. Unpublished internal report

Canadian Museum of Nature Conservation Section 2006 'Canadian Museum of Nature: 2006 Natural Heritage building risk assessment revision'. Unpublished internal report

Kwindt, M nd *Building Envelope Tip Sheet*. A report submitted to the IPM Working Group of Canadian Museum of Nature, unpublished

Waller, R 2000 'Canadian Museum of Nature (Natural Heritage Building) collection risk assessment'. Unpublished internal report submitted to Jerry Fitzgerald, Director, Collection Services Division

8

The brown carpet beetle, *Attagenus smirnovi*: how will climate change affect its future pest status?

Lise Stengaard Hansen*, Monika Åkerlund†, Terje Grøntoft‡, Morten Ryhl-Svendsen§, Anne Lisbeth Schmidt**, Jan-Erik Bergh†(deceased 2009) and Karl-Martin Vagn Jensen††

*Aarhus University, Research Centre Flakkebjerg, Slagelse, Denmark, LiseS.Hansen@agrsci.dk (author for correspondence)
†Swedish Museum of Natural History, Stockholm, Sweden, monika.akerlund@nrm.se
‡Norwegian Institute for Air Research, Kjeller, Norway, teg@nilu.no
§National Museum, Lyngby, Denmark, morten.ryhl.svendsen@natmus.dk
**National Museum, Lyngby, Denmark, anne.lisbeth.schmidt@natmus.dk
††Aarhus University, Research Centre Flakkebjerg, Slagelse, Denmark, Karl-MartinV.Jensen@agrsci.dk

Abstract

The brown carpet beetle *Attagenus smirnovi* Zhantiev (Coleoptera: Dermestidae) is an important pest found in objects of organic origin in museums of cultural and natural history in Europe. This paper presents the results of a project 'Insect Pests and Climate Change – the *Attagenus smirnovi* project' (http://smirnovi.natmus.dk/index.html) in which four partner organisations collaborated in an attempt to predict how the pest status of this species will be affected by future climate changes.

New data were collected on its distribution in Europe. It has now been reported in many European countries. Laboratory experiments were conducted to describe the effect of temperature (20, 24 and 28°C) and relative humidity (50 and 75 per cent relative humidity) on how much this pest consumes of different materials (wool, old and new skin, and feathers). The project included laboratory investigations on how much this pest consumes of four different organic materials in relation to temperature and relative humidity. Consumption generally increased with increasing temperature. Wool was consumed in the greatest amounts and was thus most damaged. The average consumption of wool during 90 days at 28°C was 5mg per larva, which was about 50 per cent more than the consumption of leather. Survival on wool, however, was poor; only about 50 per cent of the larvae survived on wool alone for three months. No mortality was seen when they were maintained on skin.

Model calculations were conducted to estimate the effect of predicted climate changes on future problems with this pest. It is probable that increasing temperatures in the Nordic area will increase the indoor reproduction and food consumption capability in this century. Increased spread of *Attagenus smirnovi* by flying is most probable in the south and south-eastern part of the Nordic area where the number of sun-hours during summer are expected to increase.

Keywords

Brown carpet beetle, *Attagenus smirnovi,* pest, museums, consumption, distribution, temperature, climate change

Fig 1
Larva of Attagenus
smirnovi *(Coleoptera:*
Dermestidae).
(Photo: R. Fortuna)

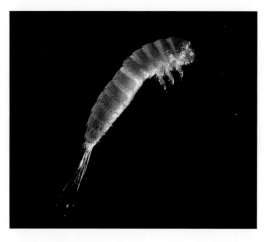

Fig 1
Larva of Attagenus smirnovi *(Coleoptera: Dermestidae).*
(Photo: R. Fortuna)

Introduction

The brown carpet beetle *Attagenus smirnovi* (Fig 1) and other species of dermestid beetles are important pests found in objects of organic origin (such as skin or wool) in museums of cultural and natural history in Europe (Pinniger 2001). *A. smirnovi* is the insect pest most commonly found in the National Museum of Denmark. Here, damage due to *A. smirnovi* has been found on archaeological human skin and historical animal material belonging to the Inuit collections. Skin and wool have been consumed, and hair has been dislodged.

Pest control in museums requires special care to avoid damaging sensitive and irreplaceable artefacts and cultural heritage. At present a combination of prevention, monitoring and mostly physical methods are employed to prevent development of insect infestation in museums and collections. (Ackery *et al* 2005; Bergh *et al* 2003; 2006; Strang 2009; Pinniger 2010)

Insects in general are highly affected by climate conditions, and temperature has a particularly pronounced effect on their development rate, activity, reproduction, and so forth. Within certain limits these parameters are positively correlated with temperature. Expected future climate changes are estimated to lead to an increase in temperatures in many locations around the world (Solomon *et al* 2007). This may lead to higher temperatures during the summer months and expand the warm season so it starts earlier and ends later. In museums and collections this may lead to higher indoor temperatures, most pronounced in premises without climate regulation. With higher temperatures the development of damage in museums due to insects can be expected to increase, and irreplaceable artefacts are at risk of deteriorating and being lost.

The present paper reports the results of a the project 'Insect Pests and Climate Change – the *Attagenus smirnovi* project' (http://smirnovi. natmus.dk/index.html). Four partner organisations in Scandinavia collaborated in an attempt to predict how the pest status of *A. smirnovi* will be affected by future climate changes. The project included laboratory investigations on how much this pest consumes of four different organic materials in relation to temperature and relative humidity. Data were collected on observations of *A. smirnovi* in Europe in order to construct an updated distribution map, and model calculations on future climate changes in Scandinavia were carried out to provide a prediction on the future distribution and activity of *A. smirnovi* as a museum pest.

Consumption of different materials in relation to temperature and relative humidity

Materials and methods

A laboratory investigation was conducted to determine how much material larvae of *A. smirnovi* consume at combinations of three different temperatures (20, 24 and 28°C) and two levels of relative humidity (50 and 75%RH). Large larvae of *Attagenus smirnovi* were placed in small containers (30 in each) with small amounts of one of four different materials:

- new skin (seal, <1 year old, de-haired, without epidermis)
- old skin (seal, >60 years old, de-haired, with epidermis)
- woollen material (white sheep, washed once, untreated, source: Historicum, Haderslev, Denmark)
- feathers (white, from gull *Larus* sp.)

The weight of the materials was recorded before starting the experiment. The containers were placed in controlled climate chambers under the conditions described above for three months (92 days). The weight loss of the materials and the survival of the larvae were recorded after the incubation period.

Result

The consumption of the different materials is shown in Table 1. The greatest amount consumed by a batch of larvae was 169mg of wool over the test period of three months, or approximately 6mg per larva over three months. The smallest amount consumed by 30 larvae was 30 mg of old skin.

Consumption generally increased with increasing temperature. In the case of skin, both new and old, more than twice as much was consumed at 28°C than at 20°C. The results for feathers were not reliable due to difficulties in weighing the feathers that had been severly damaged and were transformed into powder by the larvae. The amounts consumed at the two levels of relative humidity were not significantly different.

The survival of the larvae after three months was high on skin and feathers (above 90 per cent), but significantly lower on wool. Up to 42 per cent of the larvae did not survive three months on wool.

The highest average amount of wool consumed by each larva represents an amount corresponding to a hole of 1cm² in light woollen fabric. Many of the larvae on wool died during the test period, suggesting each of the surviving larvae must have consumed even greater amounts than this.

Climate conditions in museum locations where *A. smirnovi* is found

Data from the pest database at the National Museum of Denmark, Copenhagen, were used to describe climate conditions that support development and survival of *A. smirnovi*. Observations of *A. smirnovi* in the museum premises from 2001 to 2009 were extracted from the database and combined with indoor climate data from the same locations to describe the climate where *A. smirnovi* was observed.

Results

During the eight-year study period *A. smirnovi* adults and larvae were observed 153 times in rooms and collections at the National Museum, representing 22 per cent of all insects observed. The average climate in the 22 locations where *A. smirnovi* was found (±SD) was 21.2 ± 3.7°C and 43 ± 8.8%RH.

Distribution in Europe

Information was collected, among other sources, from the internet, discussion lists,

Table 1 Amounts consumed (mg ± SD) of different materials by 30 larvae of *Attagenus smirnovi* over three months at different combinations of temperature and relative humidity (RH)

(A) 50%RH

	New skin	Old skin	Wool	Feathers
20°C	57.3 ± 16.0 a	82.7 ± 63.1 a	103.3 ± 11.0 a	71.7 ± 5.5 a
24°C	89.7 ± 16.3 ab	46.7 ± 15.3 a	150.3 ± 20.0 ab	89.3 ± 13.5 a
28°C	94.0 ± 5.0 b	92.3 ± 16.2 a	164.3 ± 35.2 b	56.0 ± 18.2 a

(B) 75%RH

	New skin	Old skin	Wool	Feathers
20°C	52.3 ± 6.8 a	30.3 ± 3.8 a	103.0 ± 15.4 a	73.3 ± 7.6 a
24°C	77.3 ± 24.7 ab	59.0 ± 17.6 ab	123.3 ± 15.3 a	75.7 ± 6.7 a
28°C	101.3 ± 6.4 b	70.7 ± 17.5 b	169.0 ± 17.3 b	89.7 ± 9.3 a

Averages within the same column in each table followed by the same letters are not significantly different (P = 0.05, ANOVA and Tukey's Studentized Range Test; SAS 2002–2003).

Table 2 List of countries in Europe from which *Attagenus smirnovi* has been reported

Country	Year of first observation	Reference
Russia	1961	Zhantiev (1976)
Sweden	1962	Hagström (1981)
Denmark	1963	Hansen (1965)
Switzerland	1975	C. Huber (pers comm)
Great Britain	1978	Peacock (1979)
Finland	1979	Hämäläinen (1980)
Estonia	1981	I. Suda (pers comm)
Norway	1983	Ottesen (1985)
Czech Republic	1984	Černý (1989)
Germany	1985	Naumann (1986)
Belarus	1987	Barševskis (2001)
Poland	1999	Ruta et al (2004)
Latvia	2001	Barševskis (2001)
France	2004	Callot (pers comm)
Belgium	2006	Bruge (2008)
Lithuania	2009	A. Barševskis (pers comm)

contacts to authorities, colleagues and from the literature concerning observations of *A. smirnovi* in Europe. A total of 679 reports were collected from 16 countries (Table 2). *A. smirnovi* was first reported from Moscow, Russia in 1961 (Zhantiev 1976) and is now found in most countries north of the Alps up to north of the polar circle in Scandinavia. *A. smirnovi* has thus been observed in regions with a climate that does not support its reproduction and development out of doors (for example, northern Scandinavia). The dispersal of this pest does not seem to depend on the outdoor climate alone. In these cases the plausible explanation for the presence of the pest is that it is spread by humans, as well as artefacts that are exchanged among museums.

Future climate changes

Temperature is the factor that has the greatest impact on the development, reproduction and activity of insects. Increasing outdoor temperatures are likely to lead to higher temperatures in indoor locations with a climate similar to the outdoor conditions. For example, in buildings that may be maintained at about 20°C, but they are not cooled when temperatures rise above room temperature. An analysis of what changes in temperatures in Scandinavia can be expected as a result of climate change was conducted.

Methods

The analysis was based on data available from climate modelling performed by the Rossby Centre (www.smhi.se/en/2009). Differences in the available predictions are partly due to different models and scenarios used (IPCC A2 and B2) (Nakicenovic *et al* 2009). The climate during the period 1961–1990 was used as a reference.

Results

The results of the analyses concerning average temperature are shown in Figure 2. The climate modelling generally predicts increasing average temperatures over the next 100 years. Average summer temperatures are predicted to increase by 1 to 2°C in southern Scandinavia during the period 2011–2040, rising to an increase of 2 to 4°C by the end of the century. The largest changes are expected at the end of the century in the winter months and in the north (+6–7°C).

Other results obtained by the analyses were that over the next 100 years, the annual number of days with a maximum temperature above 20°C (about 30 days today) is predicted to increase; in the southern parts of Scandinavia it may more than double, and in the north smaller increases are predicted. Furthermore, the annual number of degree-days >20°C (sum of the daily average degrees above 20°C) is predicted to increase by 0–30 per cent in this century for most of the Nordic area, and by 30–120 per cent in the south-east.

Discussion and conclusion

The brown carpet beetle *A. smirnovi* is widely distributed in Europe and is found in areas with a climate that does not support its

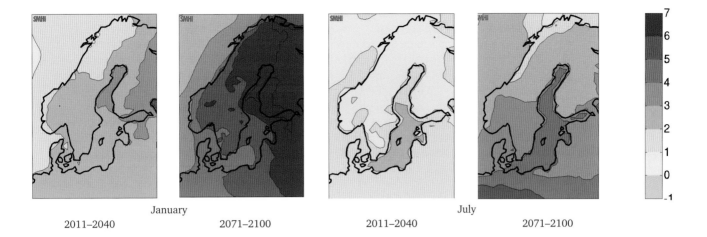

January

2011–2040   2071–2100

July

2011–2040   2071–2100

development and survival out of doors. Future climate changes in Scandinavia, with increasing temperatures predicted for especially the summer months, will most probably result in higher indoor temperatures in locations with no regulation of high temperatures.

They will also increase the number of months with climate conditions that are favourable to insects. Higher temperatures result in more rapid development of the insects and the insects can go through more generations per year. With higher consumption rates at higher temperatures, as was shown to be the case for *A. smirnovi*, damage to artefacts will be more extensive and rapid.

The present project supports concerns that climate change will affect the pest status of insects in many situations, especially in museums and collections. As *A. smirnovi* is already widely distributed it can be expected to become a major pest species in northern European museums and collections during the course of the next century.

Acknowledgements

Thanks are due to B M Pedersen and L Damberg for technical assistance. This study was supported financially by the Nordic Cultural Fund, Aarhus University, the National Museum of Denmark, the Norwegian Institute for Air Research, the Norwegian project; Climate Adaptation in Norwegian Municipalities (www.klimakommune.no) and the Swedish Museum of Natural History.

References

Ackery, P, Pinniger, D, Doyle, A and Roux, K 2005 'Heat treatment of entomological drawers using the Thermo Lignum® heat process'. *Collection Forum* **19** (1–2), 15–22

Barševskis, A 2001 'New and rare species of beetles (Insecta: Coleoptera) in the Baltic states and Belarus'. *Baltic Journal of Coleopterology* **1**, 3–18

Bergh, J-E, Jensen, K-M V, Åkerlund, M, Hansen, L S and Andrén, M 2006 'A contribution to standards for freezing as a pest control method for museums'. *Collection Forum* **21**, 117–25

Bergh, J-E, Hansen, L S, Jensen, K-M V and Nielsen, P V 2003 'The effect of anoxic treatment on the larvae of six species of dermestids (Coleoptera)'. *Journal of Applied Entomology* **127**, 317–21

Bruge, H 2008 'Attagenus smirnovi Zhantiev, 1973 (Coleoptera: Dermestidae) B. sp.n. avec information originale sur le mode de nutrition des adultes' (*Attagenus smirnovi* Zhantiev, 1973 (Coleoptera: Dermestidae) B. sp.n. with original information on adult nutrition) [In French]. *Bulletin de la Société Royale Belge d'Entomologie* **144**, 29–34

Černý, M 1989 'Faunistic records from Czechoslovakia, *Attagenus smirnovi*'. *Acta Entomologica Bohemoslov* **86**, 76

Hagström, T 1981 'Den bruna pälsängerns rätta identitet' (The true identity of the brown fur beetle) [In Swedish]. *Fauna och Flora* **76**, 141–42

Fig 2
Predicted changes in Scandinavia in monthly mean temperatures (°C) two metres above ground during January and July 2011–2040 and 2071–2100. The IPCC A2 scenario was used.

Hämäläinen, M 1980 'Attagenus smirnovi Zhantiev (Dermestidae), uusi asuntotuholainen Soumessa'. (A new indoor pest in Finland) [In Finnish]. Notulae Entomologicae 60, 230

Hansen, V 1965 'Nye danske biller 1964 (Coleoptera) (New Danish beetles (Coleoptera)) [In Danish]. Entomologiske Meddelelser 34, 123–24

http://smirnovi.natmus.dk/index.html (accessed 19 August 2011)

Nakicenovic, N, Alcamo, J, Davis, G, de Vries, H J M, Fenhann, J, Gaffin, S, Gregory, K, Grubler, A, Jung, T Y, Kram, T, La Rovere, E L, Michaelis, L, Mori, S, Morita, T, Papper, W, Pitcher, H, Price, L, Riahi, K, Roehrl, A, Rogner, H-H, Sankovski, A, Schlesinger, M, Shukla, P, Smith, S, Swart, R, van Rooijen, S, Victor, N and Dadi, Z 2000 Special Report on Emissions Scenarios. Intergovernmental Panel on Climate Change. Cambridge: Cambridge University Press

Naumann, E 1986 'Faunistische Notizen. 268. Attagenus smirnovi – eine neue Art fur die DDR?' (A new species for DDR?). Entomologische Nachrichten und Berichte 30, 270

Ottesen, P S 1985 'Attagenus smirnovi Zhantiev (Col., Dermestidae) new to Norway – a coming insect pest?' Fauna Norvegica (Ser B) 32, 108–09

Ruta, R, Konwerski, S, Kadej, M, Herrmann, A and Lason´, A 2004 'Three species of dermestid beetles (Coleoptera: Dermestidae) new to the Polish fauna with remarks on dermestids introduced to Poland'. Polskie Pismo Entomologiczne 73, 307–14

Peacock, E R 1979 'Attagenus smirnovi Zhantiev (Coleoptera: Dermestidae) a species new to Britain, with keys to the adults and larvae of British Attagenus'. Entomologist's Gazette 30, 131–6

Pinniger, D 2001 Pest Management in Museums, Archives and Historic Houses. London: Archetype Publications

Pinniger, D 2010 'Saving our heritage – pest management in museums and historic houses'. Outlooks on Pest Management 21(5), 239–41

SAS 2002–2003, 9.1, SAS Institute Inc., Cary, NC, USA

Solomon, S, Qin, D, Manning, M, Chen, Z, Marquis, M, Averyt, K B, Tignor, M and Miller H L (eds) 2007 Contribution of Working Group I to the Fourth Assessment Report of the Intergovernmental Panel on Climate Change. Cambridge (UK and New York, USA): Cambridge University Press www.ipcc.ch/publications_and_data/ar4/wg1/en/contents.html (accessed January 2011)

Strang, T and Kigawa, R 2009 'Combatting pests of cultural property'. CCI Technical Bulletin 29. Ottawa, Canada: Canadian Conservation Institutes

Zhantiev, RD 1976 'The Dermestidae of the USSR' [In Russian], Izd. Moskovsk. Univ., Moskva

www.klimakommune.no (accessed 19 August 2011)

9

Moths, Exosex™ and floor voids at Hampton Court Palace

Sam Higgs* and Jonathan Bridal†

*Historic Royal Palaces, Surrey, UK, samantha.higgs@hrp.org.uk (author for correspondence)
†Historic Royal Palaces, Surrey, UK, jonathan.bridal@hrp.org.uk
www.hrp.org.uk/hamptoncourtpalace/

ABSTRACT

This paper describes how AF pheromone traps and the registered product Exosex™ CLTab have been used together with more traditional treatment methods to identify and control high levels of activity of common clothes moth, *Tineola bisselliella*, in the Tudor State Apartments of Hampton Court Palace. A dramatic increase in common clothes moth was detected after the deployment of AF Demi-Diamond™ pheromone traps in 2009. Investigations into the source revealed that the accumulated dust in the floor voids was supporting their life cycle. To arrest the increasing number of moths, the product Exosex™ CLTab was deployed and in its first year of use it seems to have limited the increase. The use of Exosex™ is intended to allow staff time to develop and implement a long-term solution to the problem of the infestation in the floor voids.

KEYWORDS

Exosex™, common clothes moth, pheromone traps, housekeeping, floor voids

Introduction

A basic integrated pest management (IPM) programme was established at Hampton Court Palace (HCP) by the Housekeeping section in 1992. Gradually, a more comprehensive IPM programme, with better recording and analysis, improved staff knowledge and the use of up-to-date treatment and monitoring methods was developed by the Preventive Conservation section. Like many historic properties in the south-east of England, HCP has experienced a steady increase in common clothes moth activity over the past five years. The IPM programme has allowed staff to track the insect populations throughout the Palace and alerted staff to the increase in the common clothes moth in Henry VIII's Tudor Apartments. The Tudor Apartments comprise six large rooms, including Henry VIII's Great Hall and Great Watching Chamber (Fig 1). The Great Hall and Great Watching Chamber are textile-rich rooms and are lined with some of the most important tapestries in England. They are on a large scale, with over 480m² of vulnerable textile lining the walls.

Issues on the Tudor Route

Quarterly monitoring with blunder traps in the Great Hall and Great Watching Chamber had shown a gradual increase in common clothes moth since 2004, when they were first detected in this area. In 2007 there were increased sightings of moths, although the dramatic rise was a result of visual inspections by staff and was not reflected in the results from the quarterly monitoring programme.

Initially, staff attempted to ameliorate the rising population by increasing the frequency of vacuum cleaning, together with chimney sweeping

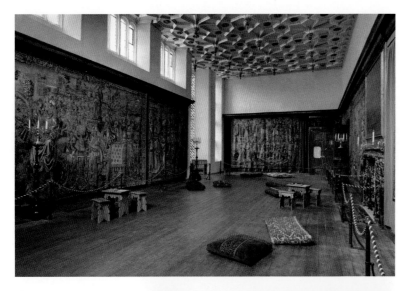

Fig 1
*Tapestries in the Great
Watching Chamber.
(© Historic Royal Palaces)*

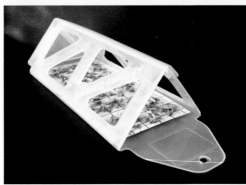

Fig 2 (right)
Demi-Diamond™ AF trap.

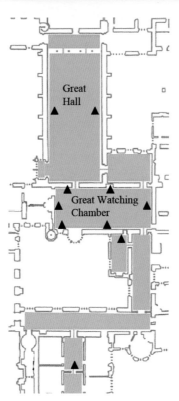

Fig 3 (right)
AF trap locations.

and localised insecticidal treatments. With these measures the number of moths found by staff both flying and settling within the rooms decreased. However, it became evident in 2009, after a trial of the highly effective AF Demi-Diamond™ pheromone trap, that contrary to the visual inspections the moth population was not decreasing and was now endemic to the room and that more direct action would be needed.

Pheromone monitoring and the results, 2009

The AF Demi-Diamond™ pheromone trap is a highly efficient trap that incorporates a slow release female clothes moth pheromone in the adhesive to attract the male of the species (Fig 2). In 2009 AF traps were trialled at Hampton Court Palace to determine their effectiveness in heavily people-populated, open-display rooms. Their success led to the project to investigate the source of moths in the Tudor Apartments. Eight traps were deployed in March and were checked and changed monthly. The decision to change them monthly rather than bi-monthly was taken after the second month as the traps were becoming heavily soiled with dust and fibres, which reduced the number of moths they could hold.

It became apparent by May that the monitoring would need to focus on one particular room, the Great Watching Chamber, where catch numbers were more than four times higher than the rest of the route. Six traps were used in an attempt to pinpoint the source of the moths within the room (Figs 3 and 4). More than 1200 adult male moths were caught on pheromone traps in the Great Watching Chamber, with over 400 of these caught in July. Despite this, the results failed to give a clear indication of where in the room the moths were concentrated. The chimney had been swept in December 2008, which eliminated this as the source. The rooms surrounding the Great Watching Chamber that are not part of the Tudor State Apartments were then thoroughly checked for signs of activity. A possible source would have been the seven tapestries and four armorials lining the walls. Although adult moths have been found on the tapestries, there was no evidence of moth damage to any of the tapestries in the Great Watching Chamber or Great Hall, many of which had been removed for treatment over the previous three years. As these possible sources were negative, other possible areas that might support moth infestation were investigated.

In an attempt to pinpoint the moth activity, an endoscope was used to look into the floor void around the edges of the room. An endoscope is a flexible optical device that can be used for inspecting inaccessible cavities, such as behind panelling. Views of the void showed that it contained a large quantity of dead skin, hair, fibres and fine grit. The dust was sampled and a microscopic inspection carried out, which found more than 30 eggs and 18 moth larvae head capsules from eight samples. This showed that the floor void was the most likely source of the moth infestation.

Floor voids and the challenges they present

The situation

Sampling from the floor void showed that there were many years' worth of accumulated dirt under the floorboards, a location that has not been accessed for a number of years. The void is approximately 15cm deep and spans the entire room, which is an immense volume of undisturbed space for moths to flourish in. The boards in this room have gaps between them varying between 1mm and 8mm where filler has been lost, which has allowed dirt and dust to collect in the void. It is this build-up of keratinaceous dust that supports the life cycle of the moth. It explains why attempts at controlling the moths through good housekeeping and localised insecticidal treatments of the accessible areas have proved ineffective. The floor in the Great Watching Chamber is late 19th century and although we have clearance from English Heritage to do minor works, it was clear that any work would involve extensive lifting of floorboards. The impact that this work might have on the collections, the building and visitor access was important in planning the next steps. It should be noted that Hampton Court Palace is open 363 days a year and Henry VIII's Tudor Apartments are the most popular attraction.

The long-term plan

A method is being devised that should control the moth population. It involves lifting every

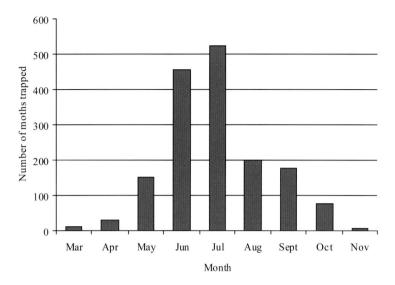

eighth row of floorboards, thoroughly vacuum-cleaning out the dust and any moth remains. By lifting every eighth board it will give us access under the floor across the entire room but this is a huge job as the floor area of the room is 225m². Desiccant dust will then be blown into the void to kill any remaining moths or larvae and should help to prevent future infestations becoming established. Possible methods of sealing the gaps between the floorboards are being investigated, to prevent future build up of dust.

During this work the tapestries will remain in position so that there is no unnecessary handling; the bottom section will be covered to protect from any dust disturbance. It is hoped that this entire project will be carried out whilst maintaining visitor access, by treating the room in sections. It will also be an opportunity to communicate to the visitors our work on controlling insects within the Palace.

However, during this planning period, blunder trap results show that the number of moths being caught year on year was increasing and the moths continued to flourish. Therefore the Exosect Exosex™ CLTab system was implemented to assist in the short term to reduce the number of moths whilst we are able to thoroughly research and implement the long-term plan to control the moths.

Exosex™

Exosex™ CLTab is a non-toxic pheromone-based product that has been developed by the company Exosect Ltd to reduce the number of

Fig 4
Moth catch in the Tudor Apartments 2009.

Fig 5
Exosex™ dispenser in position.

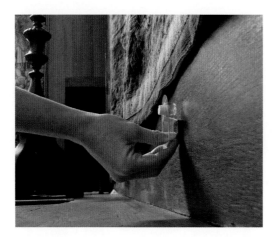

common clothes moths, *Tineola bisselliella*, without using pesticides (Fig 5). This product contains specially developed Entostat™ powder made from sustainable palm wax impregnated with the synthetic sex pheromone of the female common clothes moth. The male moths are attracted to the lure and are subsequently coated with the pheromone-impregnated powder. Male moths coated with the female pheromone powder have their senses over-loaded and are unable to locate a mate, whilst themselves attracting unsuspecting males which in turn pick up the powder. The resulting confusion of mating activities leads many females to remain unmated and produce non-viable eggs, and this results in a reduced moth population (Exosect website 2011).

Product of this type were developed for the agriculture (and later the food industry), where pests such as codling moths and Indian meal moths cause serious damage. The clothes moth version, Exosex™ CLM, was launched in 2005 (Exosect News Archive 2005). Developments in the product have seen a change from a loose powder distributed in the bottom of a large tray, similar to the large hanging traps used to monitor clothes moth, to a small tablet placed in a discreet clear dispenser. This makes it a more suitable product to use in a historic house setting. Additionally, the discreet clear dispenser means there is less chance that the visitor will notice the product and tamper with or remove it.

Successful trials of the Exosex clothes moth product at the Royal Opera House, London were followed by a trial at Marble Hill House, Twickenham (Lauder 2010). This demonstrated that Exosex™ can limit moth populations and keep them at a low level.

Using Exosex™ at Hampton Court Palace

There are several reasons for choosing Exosex™ to assist with controlling the moths at Hampton Court Palace. It is a product that could easily be managed in-house by our preventive conservation team. It is non-toxic, which is good for the staff deploying the product, and also for inquisitive visitors. The objects can remain on display, which is important due to the scale of work in removing seven tapestries and the unnecessary handling involved. The rooms can remain open to visitors, who rarely notice the dispensers. Most importantly, the product would reduce the numbers of moths whilst we researched and implemented a long-term solution.

Deployment method

With the conservation team's extensive IPM knowledge, we were able to deploy the product and collate the trap data in-house. The efficacy of Exosex™ was monitored using the established AF traps and comparing the data to the previous year. The product was first deployed at the beginning of June 2010, which was later than we intended as the moths begin to emerge in March at Hampton Court Palace. The discreet tablet dispensers were placed every five metres around the perimeter of the room at a height of around two metres. Where this was not possible, they were placed along the skirting, attached using double-sided very high-temperature masking tape, a tape that is coated with a low tack thermo-stable silicone adhesive. A tablet was also placed above each doorway. This decision was taken as we have found moths throughout Hampton Court Palace resting at high levels, in some cases as high as five metres above the floor. The deployment of Exosex™ was also extended to the adjoining room, the Great Hall, where there was also moth activity, with a total of 30 dispensers spread across the two rooms (Fig 6).

Dispensers were changed every two months and then removed in October. The decision to remove them in October was based on the AF trap results from 2009 and 2010 (Fig 7). In October 2009 the trap catches dwindled and the results from September 2010 were similar.

Results

It was predicted that, within the first two–three months of deployment, there could be an increase in our catches. This is due to the male moths being drawn out from their hiding places by the pheromone in the tablets. During the June to October deployment period there were indeed many more reported sightings of moths than in previous years. When comparing the 2009 and 2010 results, this flushing out effect can be seen, coupled with a decrease in moth numbers by two-thirds in September and three-quarters by October (*see* Fig 7). These results are encouraging despite the delay in deploying the product until June.

Conclusion

As a result of the initial success of the treatment and the ease of managing the product in-house, the product will continue to be used during 2011. Deployment will start at the beginning of the moth life cycle in March, and it is hoped that this will give better results in reducing the number of moths whilst the long-term plan is developed.

It is probable that the long-term plan will be implemented during the early part of 2012 and the use of Exosex™ will continue that year to remove any remaining adult moths from the breeding cycle.

As with many insect control methods, Exosex™ is not a total solution, but an additional weapon in the conservator's arsenal against the constant threat of an infestation.

Acknowledgements

The authors would like to thank Nicola Hugget and Andy Thompson from Exosect, for their expertise in deploying the product at Hampton Court Palace; and David Pinniger for his continued support and advice.

References

Exosect News Archive 4 October 2005 *Exosect Hits Right Note With The Successful Launch of Exosex™ CLM*. www.exosect.com/media_news/documents/ExosectHitsTheRightNoteWithTheLaunchExosexCLM04-10-05.

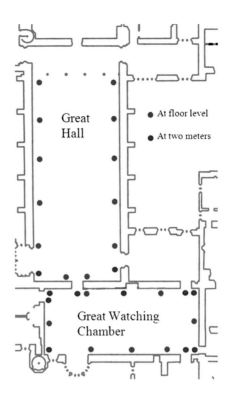

Fig 6
Exosex™ locations.

Great Hall

● At floor level
● At two meters

Great Watching Chamber

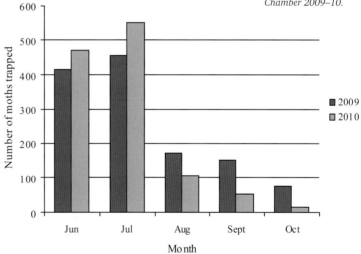

Fig 7 (below)
Moth catch in the Great Hall and Great Watching Chamber 2009–10.

asp?css=1&pg=9 (accessed 21 January 2011)

Exosect website 2011 *Exosex™ CLTab: Pesticide free, pheromone based trap enhancer for clothes moth.* www.exosect.com/pest_solutions/exosex_cltab/exosex_cltab.asp?css=1 (accessed 21 January 2011)

Lauder, D 2010 The Exosex™ CL Moth Confusion System and English Heritage 2010. Poster, London: English Heritage www.english-heritage.org.uk/content/imported-docs/a-e/exosex_system_EH_ipm_poster.pdf (accessed 27 June 2011)

10

The wider use and interpretation of insect monitoring traps

Robert Child

Conservation Consultant, Cardiff, UK, Bob.child@historyonics.com

ABSTRACT

The use of insect traps has been a central part of integrated pest management (IPM) in cultural institutions for many years. With the increased variety of traps available, and more accessible information resources on identification of trap catches, more and better information is available to the collections manager. A better interpretation of the trap catch can lead to a faster and more accurate response, which can lead to fewer eradication treatments and less damage to the collection. The future will lie in better training of museum staff in IPM with a more focused approach to insect identification, and trap management and interpretation.

KEYWORDS

Insect pests, insect traps, insect monitoring

Introduction

Integrated pest management (IPM) strategies have evolved rapidly since the late 1980s. Before then the emphasis in museums was on prevention and prophylactic treatments such as regular fumigation of premises and routine pesticide treatment of incoming objects and its packaging (Story 1985). In 1986 the introduction of the Control of Pesticides Regulations in the United Kingdom (UK), and similar legislation in other countries, forced a re-think in the management and control of insect pests in cultural institutions. The use of sticky insect blunder traps for the early detection and monitoring of insect pests became routine in many institutions in the last 20 years and now has become an integral part of the preventive conservation programmes of many museums, libraries, archives and historic properties throughout the world, (Child and Pinniger 1994; Pinniger *et al* 1998).

The use of non-attractant blunder traps that rely solely on a sticky trap surface has revolutionised pest management strategies, in that it reverses the previous philosophy of using prophylactic measures such as regular use of pesticides on collections and in collection areas. These measures carried out in the past have given rise to unacceptable toxic residues and unnecessary treatments. Insect blunder traps, and those enhanced by attractants, give early recognition of an insect infestation, identification of the culprit and an indication of a suitable treatment before irreversible damage is done. Furthermore, the traps give additional information on the integrity of the space in which they are set out. Is it dirty, dusty? Are other outdoor creatures coming in? Is the area well maintained? Insect blunder traps provide an inexpensive, low labour input method for monitoring not only the presence of insect pests, but the workings of the store or gallery space itself (National Park Service 1998).

Types of traps

Blunder traps

These are traps that catch insects, which die on the traps, but are not attracted by any lure such as light, scents, food and so forth. They are usually small cardboard traps with a sticky glue interior, which insects blunder into and are caught and subsequently die. There is also a flat plastic trap that holds a sticky board insert that can be very useful in some situations.

The sticky content of the glue is usually based on an inert organic compound such as poly-isobutene. This is sometimes modified with an equally inert microcrystalline wax. Both are then applied to a cardboard container which forms the actual traps.

The design of the trap is fundamental to the success of the catch and type of insect it will collect. Experience has shown that wedge shaped traps work best to catch both flying and crawling insects; but if the vertical height of the trap is less than 20mm, flying insects such as clothes moths are not so readily caught (Fig 1).

Placing of traps

Insects are naturally secretive and tend to inhabit quiet, undisturbed areas. Blunder traps are best placed on the floor, tight up against skirtings. They should ideally be placed in wall corners, next to doors and particularly near fireplaces/chimney areas: birds' nests in chimney flues are a source of many insect pests. There is no recommended specific distance between traps, although some suppliers suggest placing them at 2–3m intervals. Successful insect monitoring requires considerable experience but one of the guiding principles is to only set out as many traps as you can realistically manage. In warmer conditions (above 20°C) many insects can fly, so hanging traps at various heights can be very effective (Killgerm 2011).

Interpretation of the trap catches

Insects and associated animals such as spiders and woodlice caught in blunder traps can give

a wealth of useful information about the site and how it is being managed (Pinniger 2001). These include:

- the presence of insect pests – note that empty traps may not necessarily signify no insects present but rather that the trap is misplaced
- identification of the pest type, which will indicate what objects/materials are at risk
- the presence of larvae, such as carpet beetle larvae (woolly-bears), which indicates that the infestation is close by and almost certainly internal
- whether the adult beetles and moths have come from an internal or external source. Insects such as furniture beetles should be examined to see if they have flown (having done so they cannot always neatly refold their wings under the covering wing cases)
- the position of insects on a trap, which can indicate the source of the infestation. Thus, if all insects are caught on one side of the traps, further investigation on this side is called for. Traps placed in a grid pattern can, by the various numbers caught, indicate infestation sources
- an indication of local high humidity – some types of insect caught, for example silverfish, booklice and plaster beetles, can suggest a local high humidity of over 70%RH
- an indication of poor sealing – non-pest animals such as woodlice and centipedes suggests poor sealing around doors and

Fig 1
Blunder trap in position at wall/floor junction.
(Photo © DBP Entomology)

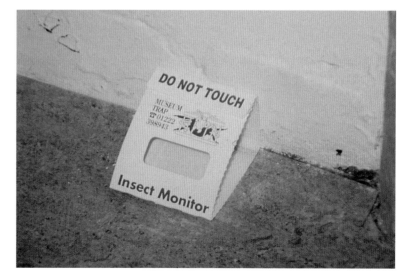

windows, which allows such animals to enter together with dirt, dust, pollution and so forth, and influence the internal environment

- the presence of dirt, dust and fluff on the trap which can indicate poor cleaning and general maintenance.

Insect traps and the law

Although ordinary blunder traps are made from non-toxic materials, they are subject to some legal constraints in the UK. These include:

- Wildlife and Countryside Act 1981
- Conservation (Natural Habitats and so forth, Regulations 1994)

The laws protect bats and other vulnerable animals such as mice that might get caught on the sticky surface (Richardson 2009). General advice is that if there are bats in or near the area being trapped:

- use a style of trap that is fully enclosed and that a bat cannot enter (entry gaps of less than 15mm)
- do not set traps in bat roosts such as a roof space (unless licensed to do so) or adjacent to a roost.

Neither the UK's Control of Pesticides Regulations 1986 (as amended 1997) nor the Biocidal Products Directive (European Union Directive 98/8/EC) consider blunder traps to be within their scope. However, sticky glue traps with an attractant are considered to be within the scope of the Biocidal Products Directive. Owing to their non-hazardous nature, sticky glue traps are not covered by the Control of Substances Hazardous to Health Regulations 1999.

Attractants

Increased trap catch can be achieved by using attractants (Kohno 1992). These come in many forms and are often specific to a few groups of insects. Providing a food attractant on a glue trap will help attract cockroaches but will not have any effect on those adult beetles and moths that do not feed. Likewise, heat will attract fleas, bedbugs and mosquitoes, but may have little effect on other pests.

The most commonly encountered attractants are described below.

Heat

The sources of heat used in traps are either a low-wattage light bulb or an exothermic chemical reaction (oxygen absorbers) and these are used together with a sticky glue trap. They are marketed for fleas and bedbugs but are also useful for attracting other insects such as biscuit beetles, *Stegobium paniceum*.

Light

Both visible and ultraviolet light emitters are used in connection with a sticky glue board or high-voltage grid type of trap. These are used routinely in high-hygiene areas to catch a wide variety of flying insects, including wasps, flies and food product insect pests (Rees 1985). They are also of value in attracting and catching cluster flies (where the glue board should be used as there is a small fire risk with the high voltage grid). UV traps will catch deathwatch beetles, *Xestobium rufovillosum*, when the ambient temperature is over 17°C, which is when they are able to fly. The colour of the traps can also have an influence on the catch, with white traps catching more deathwatch beetles than other colours (Simmonds *et al* 2001).

Food

This is normally in the form of capsules or pastes, or mixed with the sticky glue. These may be either fruit-based, which are used to attract wasps and so forth, or spice or meat-based, which are used for cockroaches. Fish meals have been used as bait for carpet beetle larvae in Insects Limited dermestid monitors.

Scents

Some insects are attracted by particular smells, thus a number of species of *Anthrenus* carpet beetle adults are attracted by the chemical p-anisaldehyde given off by certain flowers (Imai *et al* 2002). There is also evidence that deathwatch beetles are attracted to the smell of the oak-rotting fungus *Donkiaporia expansa* (Ridout and Ridout 2001).

Pheromones

These are chemicals given off by animals to modify the behaviour of other animals, usually of the same species. For instance, the bolas spider secretes a moth sex pheromone to attract its prey (Mueller 1998). Many female insects release a sex pheromone attractant to attract a mate. Most sex pheromones are species-specific, thus the pheromone produced by the webbing clothes moth, *Tineola bisselliella*, consists of two chemicals which in combination attract only this species. One of the components will attract the case-bearing clothes moth, *Tinea pellionella*, but if the other one is present, it is repellent to this species (Cox *et al* 1996). Similarly, the pheromones for different species of carpet beetle are different (Pinniger *et al* 2003). The chemicals are often complex and difficult to synthesise, and so only a few are available to put on sticky traps to increase the catch. For instance, the sex pheromone for furniture beetle, *Anobium punctatum*, is known (White and Birch 1987) and was commercially available for a few years but now is considered too difficult to produce, economically.

The fact that the EU Biocidal Products Directive states that 'products containing attractants/repellants are within the scope of the BPD' is a cause for general concern, as the high cost of registering them may mean that it is no longer economic for the manufacturing companies to continue marketing them.

Conclusion

The use of sticky glue blunder traps has increased dramatically since the 1980s and is now one of the principal tools of an IPM programme. With the recent development of graduate preventive conservation training courses, IPM is now an integral part of the housekeeping and collection care activities in most museums, libraries, archives and historic buildings. However, the need for accurate, species-level identification of insect pests is essential for a reliable recording of the changes in insect activity and for the effective use of attractants, such as the use of sex pheromones.

References

Child, RE and Pinniger, DB 1994 'Insect trapping in museums and historic houses' *in* Roy, A and Smith, P (eds) *IIC 15th International Congress Preprints*, Ottawa, 12–16 September 1994, London: International Institute for Conservation, 129–31

Cox, PD, Pinniger, DB and Mueller, DK 1996 'Monitoring populations of the webbing clothes moth, *Tineola bisselliella*, using pheromone lures' *in* Wildey, KB (ed) *2nd International Conference on Insect Pests in the Urban Environment*, Edinburgh, 7–10 July 1996, Edinburgh: ICPUE, 541–5

Health and Safety Executive 2001 *The Biocidal Products Regulations No. 880* www.legislation.gov.uk/uksi/2001/880/pdfs/uksi_20010880_en.pdf (accessed 31 August 2011)

Imai, T, Maekawa, M and Tsuchiya, S 2002 'Attractiveness of p-anisaldehyde to the varied carpet beetle *Anthrenus verbasci*'. *Applied Entomology and Zoology* **37**(4), 505–08

Killgerm Training 2011 *Insect Monitoring* No 9. Wakefield: Killgerm Chemicals Limited

Kohno, M 1993 'Response of Anobid beetle to aromatic attractants and bluelight', *in 2nd International Conference on Biodeterioration of Cultural Property Preprints*, Yokohama, Japan. 5–8 October 1992, Tokyo: The Organising Committee of ICBCP-2, 447–52

Mueller, DK 1998 *Stored Product Protection: A Period of Transition*. Indianapolis: Insects Limited Inc

National Parks Service (NPS) 1998 'Monitoring insect pests with sticky traps'. *NPS Museum Management Program* 3/7. Washington, DC: NPS

Pinniger, DB 2001 *Pest Management in Museums, Archives and Historic Houses*. London: Archetype Publications

Pinniger, DB, Blyth, V and Kingsley, H 1998 'Insect trapping: The key to pest management' *Proceedings of the 3rd Nordic Symposium on Insect Pest Control in Museums*. Stockholm, 96–107

Pinniger, DB, Child, RE and Chambers, J 2003 'Attractant pheromones of museum insect pests'. *AICCM Bulletin* **28**, 4–10

Rees, DP 1985 'Review of the response of stored product insects to light of various wavelengths, with particular reference to the design and use of light traps for monitoring'. *Tropical Science* **25**(3), 197–213

Richardson, P 2009 *Bats in Traditional*

Buildings. London: English Heritage, The National Trust and Natural England

Ridout, BV and Ridout, EA 2001 'The effect of fungi on the growth of deathwatch beetle larvae and their ability to attack oak' *in* Ridout, B (ed). *Studies in the Behaviour, Inter-relations and Management of Deathwatch Beetles in Historic Buildings.* English Heritage Research Transactions 4 Timber: The EC Woodcare Project, London: English Heritage

Simmonds, JM, Belmain, SR and Blaney, WM 2001 'Integrated pest management for the control of deathwatch beetles: Trapping' *in* Ridout, B (ed) *Studies in the Behaviour,*

Inter-relations and Management of Deathwatch Beetles in Historic Buildings. English Heritage Research Transactions 4 Timber: The EC Woodcare Project, London: English Heritage.

Story, KO 1985 *Approaches to Pest Management in Museums.* Maryland: Conservation Analytical Laboratory, Smithsonian Institute

White, PR and Birch, MC 1987 'Female sex pheromone of the common furniture beetle, *Anobium punctatum* (Coleoptera: Anobiidae): Extraction, identification, and bioassays'. *Journal of Chemical Ecology* **13**, 1695–706

11

Dealing with an infestation of *Reesa vespulae* while preparing to move to new stores

Sylviane Vaucheret* and Leona Leonard

National Museum of Ireland, Dublin, Ireland (*author for correspondence)
svaucheret@museum.ie
www.museum.ie

Abstract

An infestation of *Reesa vespulae* was discovered in the Natural History Museum, Dublin, in 2005. This parthenogenetic dermestid beetle mainly attacked our insect and mammal skin collections. We found that visual inspections were more likely to reveal the presence of activity than sticky traps and that *R. vespulae* possibly fed on the larvae of clothes moths. Freezing treatments and a thorough cleaning regime proved the most effective methods of dealing with the infestation. Our aim for the future is to prevent the transfer of *R. vespulae* to our new storage facility. Older, wooden storage units will be replaced with metal units, white foam will be used on shelving to provide a visual aid for the detection of pest activity, and where possible, specimens will be stored in plastic boxes or bagged. Quarantining of specimens will be an integral part of the move of all collections.

Keywords

Reesa vespulae, integrated pest management, infestation, insect pest, storage, quarantine

Reesa vespulae in the context of the National Museum of Ireland – Natural History

The natural history collections held by the National Museum of Ireland (NMI) are currently housed in two buildings, a kilometre apart. The first is a collections storage building where the scientific collections of zoological and geological materials are kept. This building is only accessible to curatorial staff and to visitors by appointment. The second is the Natural History Museum building, which comprises the public galleries and also holds the entomological collections.

Apart from the increased risk associated with the high numbers of visitors (on average 150,000 visitors per year), this second building, constructed in 1856, is particularly vulnerable to insect pests, due to its architectural specificities. It can be divided into three sections:

- office space separated from the rest of the building by locked doors
- public gallery on the ground floor is relatively isolated from the rest of the building, although the public circulates freely from it to the upper floors
- first floor gallery, the lower balcony and the upper balcony are all in the same open space. An attic is located above the glass ceiling. The main part of the pinned insect collection is housed in a room situated on the upper balcony and separated from the public space by a partition wall. This partition wall does not reach the level of the ceiling (Fig 1).

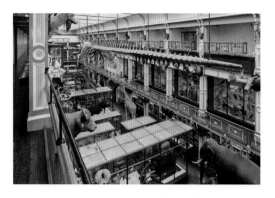

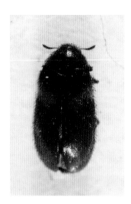

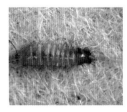

Air, dust and pests can travel freely within this third section. The roof has air vents that allow free flow of air into the attic space and is also in need of repairs. There are numerous leaks and obvious entry points for pests. The public galleries and the collection of pinned insects are therefore high priority areas for pest management.

Integrated pest management (IPM) techniques were first deployed in the National Museum of Ireland – Natural History in 2004. The trapping programme put in place, using museum sticky traps, immediately detected an infestation of the webbing clothes moth *Tineola bisselliella*. Over the following six years, this infestation was successfully controlled using a dual approach of:

• targeted trapping, using sex pheromone lures, to remove the males from the moth population and break the moths' reproductive cycle

• the establishment of a cleaning and treatment programme for both the building and the collections whereby the areas where the trap catches were the highest were prioritised.

Unfortunately, in the same period a second pest infestation was detected which proved very different and far more difficult to contain: an infestation of the dermestid beetle *R. vespulae* (Milliron 1939) (Figs 2 and 3). This second infestation was first noticed by the entomologist on the staff, Dr Jim O'Connor, during a routine inspection of the pinned insect collection in the spring of 2005.

Like several other museum pests, *R. vespulae* belongs to the Dermestidae family of beetles. It is however a more recent museum pest in Europe and is less widespread than, for instance, *Anthrenus verbasci* (Pinniger 2001). An important characteristic of *R. vespulae* is that it is a parthenogenetic species: only

females are known to occur. This makes it a particularly difficult pest to control since one individual has the capacity to start a whole population. It also means that the techniques based on sexual confusion of the males, which we successfully used to control *Tineola bisselliella*, cannot be applied to control *R. vespulae*.

In other respects, *R. vespulae* seems to have a similar life cycle to other dermestid beetles, although the literature concerned with this species is quite scarce.

What was learnt about *Reesa vespulae* during the infestation

The infestation of *R. vespulae* in our collections has allowed us to make certain observations about its behaviour in a museum context. First of all, even after the infestation became evident, the numbers of individuals of this species that were caught on sticky traps remained low. We relied mostly on the catches under the traps, rather than inside the traps, as we found the woolly-bears (larvae) were most likely to be found there. But even then, the locations where individuals were caught on the traps did not always match adequately with the areas where active infestations were discovered through visual inspections. As a result the trap-catches could not be relied on to prioritise treatments and regular, thorough visual inspections of the collections became essential.

As museum professionals, our main concern was for the collection and instinctively we first turned our focus to the specimens. Regular inspections allowed us to confirm that *R. vespulae* attacks preferably collections of pinned insects (Pinniger 2001): the main active infestations we found were on pinned insects, both in the scientific collections and on display. Amongst these, the Lepidoptera seemed to be attacked preferentially, although other groups of insects (Orthoptera), and even other arthropods (Myriapoda) were also damaged.

These collections were systematically inspected and a programme of bagging and freezing was established to treat infested insect boxes or drawers. When possible, the plastic bags were not removed after the freezing treatment to prevent re-infestations. This was not possible for the displays and repeated infestations are ongoing in this area. Indeed, the display area on the ground floor of our building proved a particularly welcoming habitat for *R. vespulae*. Pinned insects are on display in waist-height Victorian-style cabinets, below which storage space is available and is used to house our collection of mammal skins. A large number of these mammal skins were found to be infested and had to undergo the same freezing treatment as the insect collections; they were also kept bagged up afterwards (Fig 4).

Unfortunately, the display cabinets are also a few centimetres off the ground, which provides just enough space for dust to accumulate, yet not enough room to allow regular, thorough cleaning. On a rare occasion of some display cases being moved last year, newspapers from the 1940s were found amongst half-eaten lollipops, feathers and a very large amount of dirt indeed. We believe this space provides one of the reservoirs from which *R. vespulae* keeps re-infesting the insect displays.

After some woolly-bears were found on sticky traps inside some display cases containing birds, a thorough inspection of these specimens was carried out with the assistance of our Conservation Department, but no specimens were found to be infested. These observations, as well as the lessons learned from the previous clothes moth infestation, led us to shift our focus away from the collections and look for possible breeding grounds for *R. vespulae* in the fabric of the building. Many dead corners in the building had been identified during the moth infestation and these were re-explored, looking for *R. vespulae*. Two further reservoirs were identified. The first was a dead pigeon in the attic space. The bird probably flew in through one of the air vents in the roof and became trapped between roof and glass ceiling. Its dead body lay on a ventilation grille directly above the pinned insect collection on the upper balcony, where incidentally, the *R. vespulae* infestation was first identified (Fig 5). Unfortunately, the space between the roof and the glass ceiling is of extremely dangerous and restricted access. This pigeon was first noticed in April 2006 but could not be removed until August 2009, when steeplejacks were employed to do the job.

We identified the second possible reservoir from the observation that although woolly-bears could be found regularly in some display cases, none of the museum specimens themselves seemed infested. A similar conclusion

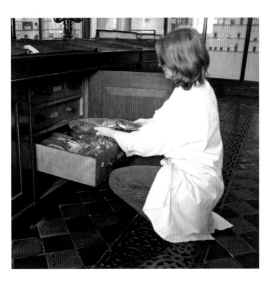

Fig 4

Insect display cases (under fabric-covers) with mammal skins in drawers underneath (kept in plastic bags after freezing treatment).
(© National Museum of Ireland; image by Nigel Monaghan)

had been reached in 2005 during the *T. bisselliella* infestation, and at the time the site of the infestation was found to be the felt lining running along the sides of each display case. The moth infestation was eradicated by painting this felt lining with insecticide.

The felt lining of display cases where woolly-bears had been found was re-inspected from March 2008 and occasionally, woolly-bear cast-off skins were found amongst the webbing of the clothes moth larvae. We were tempted to conclude from this that *R. vespulae* larvae were feeding on *T. bisselliella* individuals; however, we cannot conclude with certainty whether the dermestid larvae fed on the remains of dead moth larvae that were left after the insecticide treatments on the felt lining or whether they might have attacked live moth larvae.

The insecticide used remains active for six months but treatments were repeated about once a year, due to limited staff resources. This would have left several months each year when the insecticide in the felt, and in the bodies of the dead moth larvae themselves, would not have been present at dangerous levels. During these periods, the dermestid larvae could have fed on dead moth larvae. Also the insecticide treatments were completely interrupted

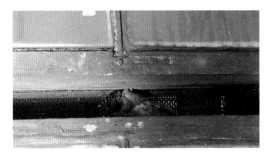

Fig 5

Dead pigeon found in the attic space. Picture taken from the attic. The entomology collection is directly under the ventilation grille.
(© National Museum of Ireland; image by Nigel Monaghan)

between July 2007 and March 2008, when the building was closed, to the public and to all staff members, due to a flight of limestone stairs collapsing.

The second scenario cannot be ruled out either. Indeed dermestid beetles of other species have been reported to attack live moth larvae (Hinton 1943). We were left wondering if *R. vespulae* might have played any active role in helping bring the *Tineola bisselliella* infestation under control.

One final observation we made about *R. vespulae* is the rarity of catching or observing adult individuals of this species. In our case only one adult was caught for every eight larvae. Also, there did not seem to be any particular seasonality in these catches, with adults caught in April and any time between June and January.

Dealing with the infestation as we move to new storage facilities

The targeted freezing treatments and the ongoing, regular inspections of our collections have been successful in limiting the damage done to our specimens and in preventing the infestation spreading to our storage building. We have not been successful in eradicating this pest completely because our museum building contains many places that can act as reservoirs. Unfortunately, plans to fully refurbish this space (Monaghan 2007) have failed to secure funding. However, the improvement of storage facilities for the National Museum of Ireland as a whole was identified as a priority in the National Development plan of 2007–13 and the museum was recently successful in securing a new building.

The Collections Resource Centre (CRC) is a former electronics plant situated in Swords, north County Dublin. The building has surrounding green areas and has been unoccupied for a number of years. It is a large site (approximately 20,000m²) with an environment conducive to clean storage of collections, due to its previous use for dust-free assembly of fine electronics. The main floor is an open-plan design, but has been fitted with a system of steel mesh cages to create separate storage areas for the substantial collections of the various museum departments to be moved.

Collections that will occupy these caged spaces include not only those from the Natural History Division, but also from the Folklife, Irish Antiquities, as well as Art and Industrial Divisions.

While every building has its own set of constraints with regards to pest management, moving to this new storage building should considerably improve the pest control situation of the natural history collections currently stored in our Natural History Museum building, particularly the research collection of pinned insects. This is because the building itself provides a better, cleaner environment and because the move of the collections will provide an opportunity to systematically freeze all specimens. Naturally, our main worry is not to bring any of our current pests with us to the new facilities, where there would be a real potential for infestations to spread to other collections (for instance, natural history collections coming from the existing storage building).

Storage solutions

Accumulation of dust and grime, exposure to fluctuating environmental conditions, inadequate physical support and security, as well as pest invasions, are the main threats to our natural history collections. Transfer to a more environmentally controllable facility will be the first step to protect the collections, as a more suitable building structure provides the first form of protection. This added physical protection should have a positive impact on the pest levels within collections.

While wooden shelving and cabinets have traditionally been used to house collections in both our current buildings, there is now a move toward implementing a system of metal shelving and cabinets in the Natural History section of the CRC. There has been some debate on the merits of steel units over wooden units in recent years. However, metal storage is a more reliable, durable alternative that has the added advantage of being more resistant to insect attack (Carter and Walker 1999; Massman 2000). Long-span shelving is already being used in parts of our current storage building and has proven to be particularly effective in housing larger specimens such as trophy game-heads and larger geological specimens, providing easier access to collections for pest inspection and research purposes. This is illustrated by our game-head room where a large number of trophy-heads is accommo-dated on open plan shelving, with white polyethylene foam on the shelves (Fig 6). The white foam provides the dual function of pro-tection against vibrations, and the white background makes dead insects and insect frass very easy to notice during routine inspec-tions of the collections. A further level of protection is provided by Tyvek dust-sheets covering the open sides of the shelving unit, preventing dust accumulating on specimens while also being breathable and waterproof (Cumberland 1993).

The area of the CRC reserved for dry natural history collections is approximately 1,500m². The plan is to fit out a portion of the space with long-span shelving to continue to accommo-date bulky collections, as well as boxed collections of smaller objects. This shelving will be standard metal beams, up to 2,500mm high, with 3 beam levels 2,300mm wide and Z-MDF shelf inserts lined with foam. The cabi-nets used will be steel with lockable double doors (Fig 7). One section of cabinets will hold geological collections in existing zero-formal-dehyde Z-MDF drawers, lined with foam, while the remainder of cabinets will accommodate mounted taxidermy and dry invertebrates. All other pinned insect collections and birds will be housed in existing wooden units in a sepa-rate area within the caged zone.

The modular system of storage implemented in the Natural History section of the CRC will comprise standard-sized storage furniture and it allows for flexibility of use. Storage boxes within this modular system fit efficiently, max-imising the space available. They will be plastic; inedible by insects, they discourage infestations and provide far better physical security for the specimens in terms of water leaks, flooding, and dust accumulation. The previous programme of treating infestations and tackling problems to stunt the spread of insect pests from one site to another has proven to be successful. The subse-quent re-housing of the collections into the systems described above should allow for a bet-ter management of pest issues.

IPM in the Collections Resource Centre, Swords

Responsibility for designing and implementing the IPM strategy for this new building lies with

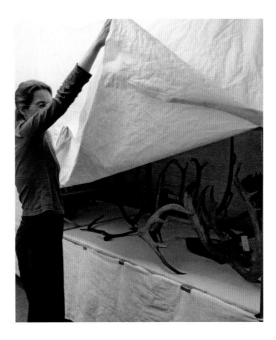

Fig 6
Long-span shelving with white foam and Tyvek curtains, for dust protection and early detection of insect infestations.
(© National Museum of Ireland; image by Rebecca O'Neill)

the staff of NMI's Conservation Department. The overall strategy will take into account many areas, including a monitoring pro-gramme to be carried out prior to moving any collections into the building, highlighting of potential problem areas and species, tackling these issues effectively prior to the move, and establishing procedures for the overall build-ing once occupied. The IPM programme is still in preparation and is not finalised in detail at this stage.

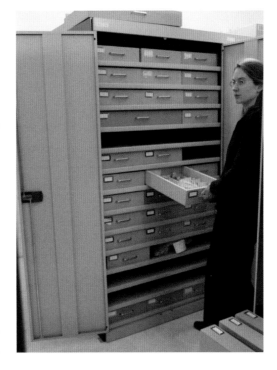

Fig 7
Example of steel cabinet, with Z-MDF drawers, used for small objects.
(© National Museum of Ireland; image by Rebecca O'Neill)

Our understanding is that the initial trapping programme carried out by the NMI Preventive Conservator involved the placement of 51 non-specific sticky traps at floor level throughout the building, with 19 of these traps containing pheromone lures (p-anis) targeting carpet beetles. A single live specimen of the carpet beetle *Anthrenus verbasci* had been discovered on a windowsill on a previous site visit by entomologist Dr Jim O'Connor. To our knowledge, trap catches to date have shown mainly non-pest species.

One of the more important considerations in the pest strategy will be the issue of quarantining collections. Considering the problems of *R. vespulae* and moth infestations in recent years in our museum building, it has always been a priority to avoid transferring these pests to our stores. We must also now consider how we can best avoid transferring these problems to the new facility in Swords. We are not only mindful of the collections of the Natural History Division, but we are also aware of the risk for other vulnerable collections that will be housed in the same space or the possibility that collections from other stores might bring their own pest populations. Our experience is that a pest population can quickly become established in the fabric of the building itself, and can subsequently become almost impossible to eradicate.

Quarantining of all specimens that enter the CRC may be necessary to ensure that we fully minimise the risk. The facility is equipped with a large, walk-in freezer that can accommodate large quantities of collections at one time. Specimens leaving our current stores will be wrapped in plastic sheeting, sealed and transported to Swords for freezing before being integrated into the collections. The Conservation Department is currently working on the overall IPM strategy for the building, and this includes the quarantining program.

Conclusion

Reesa vespulae has proved to be a challenging museum pest, but a lot was learnt while dealing with it. Its life cycle means that many traditional techniques of pest control could not be used to tackle the problem. Preventive conservation methods were found to be particularly important, such as housekeeping, regular inspections, as well as strict quarantining to avoid bringing the pest to other sites. All these methods will need to be implemented in our new Collections Resource Centre to minimise such problems in our collections in the future. The acquisition of new storage systems purposely chosen to minimise pest issues will hopefully make the new facility less prone to pest infestations than our current accommodation.

Acknowledgements

We wish to thank Karen Wilson and Rolly Read, conservators, for their input on the CRC's IPM strategy; Nigel Monaghan for his supportive and regular comments; and Patrick Butler, Rebecca O'Neill and David Pinniger for contributing pictures.

References

Carter, D J and Walker A K 1999 'Collection environment' *in* Carter, D and Walker, A K (eds) *Care and Conservation of Natural History Collections*, Oxford: Butterworth–Heinemann, 139–51

Cumberland, D R 1993 'Dust covers for open steel shelving'. *Conserve-O-Gram*, 4/2. Washington DC: National Parks Service

Hinton, N E 1943 'Natural reservoirs of some beetles of the family Dermestidae known to infest stored products, with notes on those found in spider webs'. *Proceedings of the Royal Entomological Society of London*, Series A, **18**(4–6), 33–42

Irish Government 'Transforming Ireland – A better quality of life for all'. *Ireland – National Development Plan 2007–2013*, **224**. Dublin: Government Publications

Massman, A 2000 'The wood shelving dilemma'. *Library Resources and Technical Services* **44**(4), 209–13

Monaghan, N T 2007 'The Natural History Museum Dublin, past and future'. *Museum Ireland* **17**, 48–52

Pinniger, D 2001 'New pests for old: The changing status of museum insect pests in the UK' *in* Kingsley, H, Pinniger, D, Xavier-Rowe, A, Winsor, P (eds) *Integrated Pest Management for Collections, Proceedings of 2001: A Pest Odyssey conference, London, 1–3 October 2001*, London: James & James, 9–13

12

The measured, slimline implementation of integrated pest management at the National Trust for Scotland

Mel Houston

National Trust for Scotland, Edinburgh, Scotland. mhouston@nts.org.uk
www.nts.org.uk/

ABSTRACT

The National Trust for Scotland (NTS) is a conservation charity with over 50 historic houses with collections on open display in its care. Prior to 2009, pockets of good pest monitoring practice existed, but no pest management programme was in place. In September 2008, the NTS hosted a 12-month preventive conservation internship as part of the Institute of Conservation's internship programme funded by the Heritage Lottery Fund. The focus of the placement was that the successful candidate research and draft integrated pest management (IPM) guidelines for the NTS. This paper describes the development of a programme that suits both the NTS's complexity and available resources. Risks and benefits associated with the strategy of using an intern are discussed and criteria for a property's inclusion in the first round of the programme are outlined, as is senior management's welcome, but unexpected, interest in and enthusiasm for IPM. Continuing improvement of the programme is also described, by emphasising the importance of embedding IPM training within the NTS.

KEYWORDS

National Trust for Scotland, integrated pest management, IPM, monitoring, training.

Introduction

The National Trust for Scotland (NTS) has over 50 historic houses with collections in its care. Preventive conservation is delivered through the Collections Conservation Service (CCS), principally through a team of three regionally-based conservators with a portfolio of approximately 16 or so properties each, managed by the Edinburgh-based Head of Service.

The regional conservators are variably supported by collections care officers, housekeepers, property staff and volunteers, depending on each property's staffing structure. In 2001, however, when the first Pest Odyssey conference was held, the NTS conservator function was embryonic and pest trapping within NTS properties was only carried out on an *ad hoc* basis. There were pockets of good insect pest monitoring practice through property staff taking the initiative but, almost without exception, when the instigator moved on or became absorbed in other work, the monitoring ceased.

Isobel Griffin joined the NTS in 2000 as a conservator and addressed pest management, based on the programme she had implemented at the National Museums of Scotland in the 1990s (Griffin 2001), but limited resources (time and staff) meant that there was not a consistent programme of pest trapping, recording, identification or treatment across the NTS. Some infestations of insect pests were known to the NTS conservators, but there was no information available to give an overall picture of damage being caused to the collections.

The CCS recognised the importance of integrated pest management (IPM) and, from 2008 to 2009, hosted a 12-month Institute of Conservation internship (funded by the Heritage Lottery Fund) with a focus to develop an IPM strategy for the NTS. This gave the opportunity for that incoming member of the team to take on the role of researching and writing IPM guidelines, as there was no internal capacity to do so. The IPM remit for the year-long internship was to write guidelines, but there was no expectation or agreed plan to implement those guidelines during the course of the year. The internship, for which the author was the successful candidate, started in September 2008 under the supervision of Isobel Griffin.

IPM background research

There were advantages as well as risks to having an intern write guidelines for the NTS. A major advantage was the luxury of time to devote to research, which was not possible for permanent members of the CCS staff. Research was approached in two main ways. The first was the completion of a six-week web-based course which, serendipitously, commenced within a week of the beginning of the internship and gave a focus when it was most needed. The course, 'Integrated Pest Management for Museums, Libraries and Archives', was run through the Northern States Conservation Center and required a commitment of approximately 20 hours a week study: time other members of the team could not contemplate. The modular course provided the basic tenets of IPM and, most usefully, encouraged participants to write an appropriate IPM plan for their institution as part of the coursework. This proved useful in providing a framework document from which the NTS guidelines were subsequently developed.

The second approach to research was the investigation of IPM programmes and policies at other heritage organisations. Approximately 15 IPM programmes across the sector were examined during the research phase, which highlighted another advantage an intern has: generous access afforded by conservators working in other institutions. The author is very grateful to those key members of staff in those institutions who gave their time and knowledge, access to their IPM policies and programmes and, above all, their support. Visits to conservators involved in IPM programmes in museums and heritage organisations across the UK gave an excellent understanding of how each approached IPM depending on their collections, staffing levels and the challenges posed by the buildings housing each institution's collections. It quickly became apparent that there was no single programme that matched both the needs and the resources of the NTS, so guidelines had to be tailored to fit. Understanding of the NTS's resources and capabilities were key factors in writing the guidelines, and the author was fortunate in this regard having worked as both a seasonal property assistant and a collections care officer for the NTS during the previous three years. A further factor, established when meeting conservators involved in running IPM at other institutions, was that staff training and support were central to the success of any programme, an observation which profoundly shaped the ensuing programme developed for the NTS.

Putting guidelines into practice: a pragmatic approach to IPM

IPM guidelines were completed and accepted in December 2008 (Houston 2008). It was evident, however, that guidelines were just the starting point and the next step was their implementation through a structured IPM programme led by a designated IPM co-ordinator (the intern). A protocol for setting up an IPM programme was therefore developed by the author in early 2009, which included the criteria for a property's eligibility for inclusion in the programme. A flexible timetable for sticky trap checks in the first year was proposed and, thereafter, checks would be recorded quarterly, commencing in March each year. A realistic approach was taken from the outset. The aim was not to write a protocol for the best IPM programme in the world, but one that was measured and a good fit for the resources of people and equipment available in the NTS.

A summary of criteria for inclusion in the programme and a simple protocol is detailed below.

- **Identifying a property**

  The IPM co-ordinator and regional conservator identify a property with a known or suspected problem that has a *willing* member of staff to take on responsibility for pest trap monitoring.

- **Site visit**

  The regional conservator contacts the property manager to set up a site visit at a time convenient for all (property manager, regional conservator, IPM co-ordinator, and the property staff member nominated for IPM).

- **Introduce IPM**

  The purpose of the site visit is to introduce the topic of IPM to property staff and property managers, explain why the NTS is investing resources this way, outline the role of the nominated property staff member, explain the monitoring process, and assess any potential pest problems by completing a property assessment form establishing whether the property is aware of any existing pest problems. The frequency of functions and events is recorded, together with information about any existing environmental monitoring system.

- **Initiate trapping**

  During the site visit, labelled and numbered sticky traps are placed in appropriate places and a record of trap locations is written into a log book. The nominated member of staff is given a small IPM resource kit and it is explained that checks are to be made at three-monthly intervals. After the first year, trap checks will take place in specific months: March, June, September and December.

- **Support and monitor progress**

  The IPM co-ordinator (and regional conservator if available) makes a return visit after three months to check all the traps and record results with the IPM nominated member of staff. The role of the IPM co-ordinator is to support this key member of staff as well as to ensure that trap finds are recorded accurately and to answer any questions.

- **Ongoing support**

  In the first year, the IPM co-ordinator returns at the six-month and nine-month mark to support the IPM nominated staff member in recording trap finds. In some circumstances these additional visits are unnecessary, but it is for the regional conservator to decide whether the nominated member of staff needs the extra support. If in doubt, err on the side of caution, as it is vital that these key members of staff are confident and accurate with their pest identification.

- **Review and trap data collection**

  At the twelve-month mark, the IPM co-ordinator and regional conservator visit to record the latest finds and use the year's data to produce a short annual report that records the pest species trapped and levels of infestation, records any treatments and makes any recommendations necessary to combat ongoing pest problems.

Commitment, enthusiasm and a shoestring

It is understood by all in the CCS team that the success of the NTS's IPM programme lies with the property staff who take on the responsibility for monitoring pest traps in addition to their demanding property duties. Likewise, the regional conservators with their portfolios of 16 or so properties are essential in fulfilling the obligations of trap checking at every property in the programme four times a year. It was important therefore to ensure that everyone had equal access to resources for their part in the programme. For example, some staff (depending on their roles) had access to the NTS's computer systems, but many did not and the trapping logs therefore had to be paper-based. A paper-based system was also seen as a way of keeping the programme as simple as possible, with less likelihood of loss of data through human error or catastrophic computer failure. It is envisaged that, as investment in information technology (IT) by the NTS increases, more staff will have access to the Trust's computers, and so electronic versions of the IPM programme's log books have been created with this in mind. The regional conservators, however, do have access to these electronic versions, and update them annually when writing the annual IPM reports.

The resource kit (Fig 1) given to each property-based nominated IPM staff member,

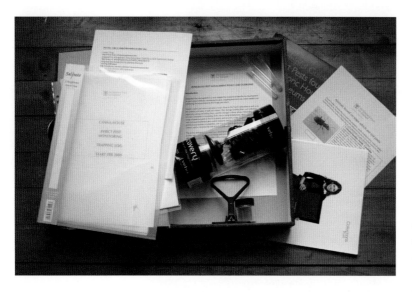

Fig 1
IPM resource kit for
property staff.
(© National Trust for
Scotland)

was high that the implementation of any programme would collapse after the completion of the internship due to the loss of a central co-ordinator. This risk was, however, identified early on and is reflected in the IPM strategy, as it was agreed essential that the roll-out of any IPM programme for the NTS would be phased at a pace that the CCS team could maintain and support.

With the pilot phase of three properties successfully underway in early 2009, a phased roll-out of the programme could then be drawn up by the co-ordinator/intern. The phasing was designed so each regional conservator would have an equal share of the implementation of the programme on an annual basis, approximately three properties per year. Properties were added at a sustainable rate to ensure that nominated property staff could receive the support they needed and, by the end of 2009, eleven properties were participating. Nine more were added in 2010, following the appointment of a fixed-term conservator whose responsibilities included acting as the IPM co-ordinator.

consists of a box file containing: documents written by the IPM co-ordinator (*Introduction to IPM*), the NTS's IPM guidelines, a resource sheet (a list of useful websites prepared by the IPM co-ordinator), the trapping log book, a supply of sticky traps, a magnifying glass, test tubes and the English Heritage/Collections Trust pest identification poster. Subsequently, each property's resource kit has had added to it a copy of *Pest Management: A Practical Guide*, (Pinniger 2008) and a digital microscope. Any IPM equipment, for example, traps, chemicals, reference material or viewing aids, are funded by a central budget, and not from the property's budget, as it is considered important to demonstrate to property managers that IPM is not a financial burden on property resources. The financial burden is not, therefore, an issue as to whether a property will commit to the IPM programme.

Pilots and phasing

In December 2008, the pilot phase of the programme began with three selected properties chosen because each had the support of a collections care officer or an experienced housekeeper as the nominated IPM member of staff. The aim of this pilot phase was to test the property assessment protocol, as well as the documentation and recording systems with property staff, and to act on any feedback before the programme was rolled out NTS-wide. And, it is at this juncture that the disadvantage of using an intern to develop an IPM strategy became all too clear, as the risk

Training

The key to the success of the NTS programme is first to identify and then develop staff who are keen to be involved, as the NTS cannot manage an IPM programme without the willing participation of these colleagues. The co-ordinator is responsible for ensuring nominated staff have competence as well as confidence, and training starts with the initial visit. IPM is explained and basic information is given, and the co-ordinator discusses with the nominated member of staff his or her IPM knowledge and assesses their confidence at identifying insect pests. When the co-ordinator returns for the quarterly trap checks, nominated staff are encouraged to use the resources in the IPM kit to aid correct identification.

With three pilot properties underway in early 2009, the IPM programme was introduced to a wider audience in the NTS though inclusion in the Collections Care workshops which take place at three venues across Scotland in February each year. These workshops are the vehicle by which the Collections Conservation team delivers hands-on training and raises awareness of the care of the NTS's collections to a wide cross-section of roles

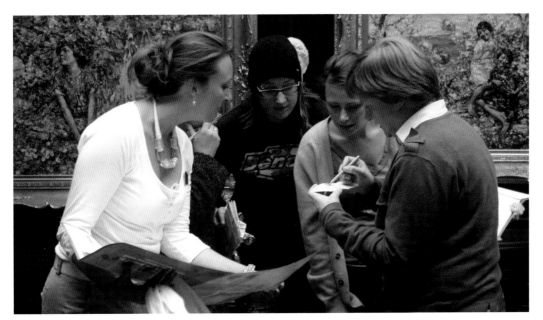

Fig 2
Staff training in pest
identification, Broughton
House.
(© National Trust for
Scotland)

across the organisation. Two sessions in the 2009 workshop were delivered: the first gave participants an introduction to IPM, and the second was a basic practical pest identification session (Fig 2). A session on advanced pest identification was included in the 2010 Collections Care workshops.

A major training initiative took place in November 2009 when David Pinniger led a pest identification workshop for NTS nominated staff. David's presence added authority to the programme and raised the profile of the members of staff involved, colleagues who do not always receive the recognition and credit they deserve. The workshop increased everyone's confidence in pest identification and, through David's leadership, re-emphasised that pest identification is enjoyable. The author subsequently has helped introduce a structured IPM training scheme to the NTS, which David Pinniger continues to support, and it is planned that this will be an annual two-day training event. For instance, in November 2011, the first day will be an introduction to IPM for those new to the programme or requiring a refresher session, and the second day will be an advanced pest identification day for staff that have a basic knowledge of IPM. Staff attendance at these training days has been incorporated into the protocol for adding properties to the IPM programme and attendance is noted in the IPM annual report for a property.

Nominated staff are encouraged to take responsibility for IPM at the property where they work, and comments and feedback on the resources, documentation and solutions to problems are positively received. Bat-proofing traps are a good example of staff involvement: a number of NTS properties have bat roosts, and bat-proofing is essential to comply with legislation to protect the bats. One enterprising member of staff devised stainless steel cages using drain-pipe balloons to protect sticky traps, and these have been tested at other properties and modifications have been made in response to staff comments. Similarly, trapping log sheets have improved significantly with the incorporation of work by enthusiastic property staff that had produced their own documentation prior to IPM being introduced across the NTS. Training has also been greatly enhanced by the introduction of low-cost digital microscopes, first introduced in 2009 for use by the three regional conservators to assist with pest identification. The microscopes, which proved easy to use and capable of recording acceptable quality images and videos, have since been trialled with staff at training days (Fig 3) and proved so successful that microscopes were purchased for each property's IPM resource kit. These microscopes are used by staff to assist with pest identification on site, and images of insects which are difficult to identify can be captured and sent for further identification to the IPM co-ordinator. The digital microscopes have also proved of interest to other departments, such as the NTS's Learning and Archaeology Services.

Fig 3
Training using digital
microscopes.
(© National Trust for
Scotland)

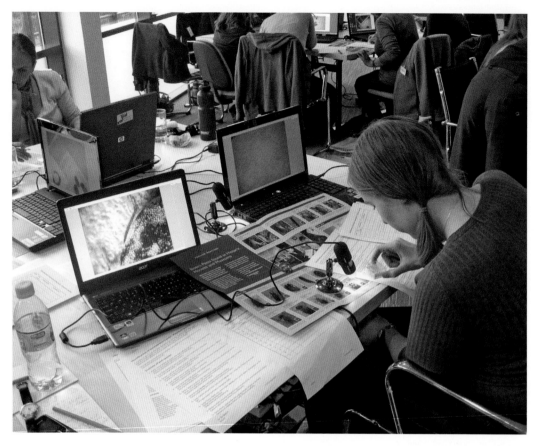

Buy-in

Through evaluation of IPM training and speaking with property staff, it is apparent that nominated staff enjoy and understand the value of the work for which they are responsible. The profile of IPM in the NTS has steadily increased since the programme began in 2009 and, by running IPM training in consecutive years at the annual Collections Care workshops, the message of pest management has reached a wide NTS audience, including: property managers, learning officers, catering staff, guides, volunteers, marketing and development managers, as well as directors.

There has also been unexpected buy-in from senior management, who use IPM in both internal and external presentations to demonstrate how the NTS is working to care for its collections and, of course, the cost–benefit point can be made very neatly. This has raised the profile of IPM within the NTS which, in turn, has increased the general awareness of what the Collections Conservation Service does to protect and conserve the NTS's collections. IPM has immediacy: damage can be seen, pests can be watched at terrifying magnification with digital microscope video footage, and treatments can be quickly initiated and mitigation achieved.

Findings from trapping

From the analysis of IPM data, it is clear that the NTS has major insect pest infestations in collections at most of the historic properties on the programme. This may be unsurprising as one of the criteria for inclusion in the programme is the existence of a known or suspected pest problem. The principal recurring species are webbing clothes moth, *Tineola bisselliella*, and furniture beetle, *Anobium punctatum*. Now that this knowledge exists, all property staff can be alerted to the signs of new infestations of *likely* pests and regional treatment programmes can be planned (for example the purchase and placement of a scientific freezer for a group of properties at an appropriate location). It is intended that the data collected will be incorporated into a national pest database and it is recognised that, with its Scotland-wide remit, the NTS has a unique part to play.

Conclusions

The introduction of IPM guidelines, supported by a phased, measured, sustainable and low-cost programme of property inclusion, has proved successful. Pest infestations have been identified and mitigation procedures initiated. At the centre of the programme are the nominated members of staff at properties, who carry out the pest monitoring and record trap finds and alert their regional conservator to the presence of pests in the collections.

Nominated staff are supported by a continuous training programme, annual pest identification workshops, visits from the IPM co-ordinator and regional conservators, as well as the provision of essential resources to encourage correct and confident pest identification. By using property staff and low-cost equipment, it has been possible to implement a cost-effective IPM programme that suits the resources of the National Trust for Scotland. The programme will continue to expand due to the establishment of a permanent central preventive conservator post, whose responsibilities include IPM co-ordination, and the management of the phased introduction of further properties into the scheme at a pace dictated by the resources available.

Acknowledgements

The author would like to thank Isobel Griffin for her supervision and advice, and David Pinniger for his support of the programme and help with training. Many thanks to Clare Meredith, Head of NTS Collections Conservation Services, for her continued support and help with editing this paper.

References

Griffin, I 2001 'The development of an integrated pest management policy for the National Museums of Scotland', *in* Kingsley, H *et al* (eds) *Integrated Pest Management for Collections, Proceedings of 2001: A Pest Odyssey Conference,* London, 1–3 October 2001, London: James and James, 21–7

Houston, M 2008 'National Trust for Scotland: Integrated Pest Management Guidelines', unpublished internal NTS document

Pinniger, D B 2008 *Pest Management: A Practical Guide*. Cambridge: Collections Trust

13

Driven to distraction by moths: IPM on the Riverside Museum Project

Gretel Evans

Gretel.Evans@gmail.com

ABSTRACT

This paper describes the pest management work carried out on the transport and technology collection during the Riverside Project, a capital funded project to build a new museum to re-display the collection in Glasgow. Monitoring and treatment of the collection, together with the results, are discussed. The challenges and mitigating actions of the integrated pest management strategy for the new Riverside Museum are highlighted; the importance of the basics of pre-planning and good housekeeping is stressed.

KEYWORDS

Common webbing clothes moth, infestation, transport collection, pest infestation treatment, risk rating strategy, innovative displays, good housekeeping

Introduction

The Riverside Museum is a £74 million capital project funded by Glasgow City Council, the Heritage Lottery Fund and the Riverside Museum Appeal. As a world-class visitor attraction it displays part of Glasgow Museums' internationally and nationally significant transport and technology collection, previously housed in Kelvinhall, Glasgow. The transport and technology collection has suffered from a long-standing infestation of common webbing clothes moth. The Riverside Museum Project provided the resources and impetus to implement an integrated pest management (IPM) programme, the focus of which was to tackle the severe clothes moth infestation.

Background

Displayed at the Museum of Transport, Kelvinhall for over 20 years, the transport and technology collection, as well as the natural history and entomology collections in store at the old museum, have suffered long-standing and severe infestation and damage by common webbing clothes moth, *Tineola bisselliella*. Much of the transport and technology collection has been re-displayed in the new Riverside Museum, together with a substantial social history component. The iconic building was designed by Zaha Hadid Architects and the highly innovative displays worked up by museum designers, Event Communications.

The building itself and the object displays present challenges with regards to pest management. Key attractions include 31 bikes encircling an infinite velodrome which hangs from the ceiling; a display of 30 cars mounted on a wall in three rows up to a height of nine metres; a further seven cars are mounted on one continuous ramp to a height of over five metres. Story displays follow the current trend for open display of objects, and cases where they exist are frequently no more than dust-covers. The museum design presents inaccessible dead spaces under plinths and display furniture. These issues need to be mitigated as far as possible by a comprehensive pest management strategy.

Museum of Transport, Kelvinhall

The problem of common webbing clothes moth in the Museum of Transport, Kelvinhall had been evident for over 10 years, with major infestations in natural history specimens in store, and within a wide variety of vehicles on display.

Suspected sources of the moth infestation vary from museum acquisitions, such as a recently inhabited caravan, to objects on loan from other museums, through to a dead pigeon. One or more of these may or may not be the source of the infestation. However, the possible sources of the infestation do provide lessons to be learnt. Quarantine procedures and good housekeeping regimes are important aspects of an integrated pest management programme.

Common webbing clothes moth

The common webbing clothes moth, *T. bisselliella,* is small, fawn-coloured with a brush of ginger hairs on the head. It dislikes light, preferring dark undisturbed areas and feeds mainly on keratin-based materials. This makes fur, feathers and wool a favourite, but by no means the only food. In nature it would normally complete its life cycle over the period of a year. However, with the higher temperatures in the museum it is not unusual that there is more than one generation per year.

IPM programme

Glasgow Museums has had an IPM programme in place at the Museum of Transport, Kelvinhall since early 2006. The programme consisted of general monitoring for museum pests together with an in-depth focus on monitoring for, and treatment of, common webbing clothes moth infestations.

Sticky blunder traps placed in the galleries and stores would catch silverfish (*Lepisma saccharina*), hide beetle (*Dermestes maculates*), spider beetles (*Ptinus* sp.) and booklice (*Liposcelis* sp.), but not in great numbers or areas of concern. With the exception of hide beetle

infestations found in bins around the picnic area there were no other pest problems of concern in the old museum.

Several catches of common webbing clothes moth were recorded on the sticky blunder traps. However, the number of catches was not indicative of the severe infestation that was known to be within the collection. Evidence of the moth infestation was seen in damage to the collection, and sightings of adult moths by museum staff.

Decant

In conjunction with the Riverside Project the stores within the Museum of Transport were decanted to new storage facilities at Glasgow Museums Resource Centre (GMRC). A team of technicians packed all objects for transportation to GMRC. They were trained to recognise signs of pest damage and highlighted objects of concern to the conservation team who carried out treatment as necessary. Objects brought to attention were mainly associated with webbing clothes moth damage, but these were isolated incidences. There was no indication of a major infestation within the collection in storage.

Quarantine

A quarantine strategy was developed as part of the IPM programme. Over 3000 objects were to be displayed in the new museum building. Some were housed within the Museum of Transport but many were stored elsewhere. Quarantine was all the more important due to the large numbers of objects entering the old museum at Kelvinhall to undergo conservation for re-display. All objects, along with props, entering the museum were subject to quarantine. Ideally all materials including educational resources, shop stock, and vegetation should be subject to quarantine or treatment.

The vehicle collection

The main focus of pest management at the old Museum of Transport was on the severe infestation of the vehicle collection by common webbing clothes moth. As there were nearly 150 vehicles on display, a multi-disciplinary team was formed, made up of transport, textile, and object conservators, led by the conservator, with responsibility for pest management. Regular meetings kept the team briefed on developments and progress within the vehicle pest management programme. In addition, all museum staff, including contract cleaners, were briefed on the pest management programme.

Assessing the problem

Before tackling the problem, an assessment was needed of the extent of the infestation within the vehicle collection. Blunder sticky traps with pheromone lures were laid in every vehicle that was enclosed and relatively well sealed, but not in vehicles such as open-top cars, windowless carriages, or those with large gaps to the outside. The traps were checked every month and the number of moths caught recorded on a spreadsheet for each vehicle.

Surveying the vehicles

To supplement the monitoring of the transport collection a thorough survey of each vehicle was carried out. Primarily, signs of moth infestation were looked for in the form of live or dead moths, webbing, larvae, eggs or damage. Where signs of the infestation were visible, it could be established that a number of vehicles were actively infested.

Fibre identification

In addition to the visual survey, all textile, upholstery and padding materials within vehicles were sampled, taking a few fibres to allow their identification by microscopy. The textile conservator identified whether fibres were plant, animal or synthetic although frequently identifications were more detailed. Fibre identification showed that a variety of different materials were used over the years, including wool, cotton, jute, horsehair, bast, and synthetics.

Vehicle vulnerability rating

Surveying the vehicles, along with the fibre identification, meant it could be established straight away if a vehicle was likely to be vulnerable to infestation or if the infestation was visible. Monitoring each vehicle over a period of time with pheromone lures helped to clarify the situation. If there were no moths caught on a pheromone trap over a period of a year, then it could definitely be said that the vehicle was not infested.

By combining all this information, each vehicle could be placed into a vulnerability category which denoted not only whether the vehicle was currently infested, but how liable it would be to infestation. These categories then informed the frequency of monitoring. For example, those cars in the highest category, vulnerability group one were actively infested and checked monthly; whereas those with no vulnerability did not need to be monitored. Vehicles that contained vulnerable materials, but no infestation were monitored every three months.

Allocation of vehicles to vulnerability groups allowed targeting of resources: it eliminated the need to check vehicles that do not present a risk enabling resources to be concentrated on those that pose a problem.

Treatment

The same principles, as with objects, were applied to select a treatment for infested vehicles. Several factors guided the choice: a short treatment time is advantageous; relatively inexpensive treatments are favoured as resources are finite; ideally the treatment could be carried out in-house by conservators or technicians. Low-temperature treatment, the application of a suitable conservation-tested insecticide, or a combination of the two were considered.

Treatment of a vehicle varies according to the type of vehicle and the location of the infestation within. Where possible, if size and material of the item to be treated allows, the preferred choice is low-temperature treatment, using an ultra-low temperature chest freezer which reaches −30°C, as required for a 72-hour treatment. In relation to vehicles, items treated by this method include removable carpets, mats and seat pads. Where a seat pad is incorporated into a metal frame this is removed as well and the whole unit frozen. One of the major benefits of low-temperature treatment is the ability to reach inaccessible areas such as horsehair stuffing within seat pads.

When low-temperature treatment is not possible, it is usually due to the size of an item or because it is fixed in place. In these cases a conservation-tested, water-based insecticide is applied. This method is used to treat fitted

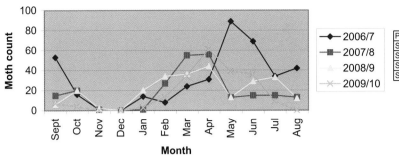

Fig 1
Monitoring results for
webbing clothes moth in the
small vehicles group.

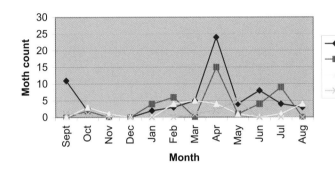

Fig 2
Monitoring results for
webbing clothes moth in the
wooden vehicles group.

carpets and upholstery within vehicles, as well as head-linings and any other non-removable fabrics that require treatment.

Results

Monitoring of the vehicle collection began in September 2006. For ease of monitoring, the vehicle collection was split into three groups. The largest in number of these groups is the small vehicles, comprising mainly cars; next is the large vehicle group, encompassing trams, buses and subway carriages; the smallest of the groups is the wooden vehicles, all of which are horse-drawn carriages.

Results were recorded on a spreadsheet, with numbers of moth per vehicle noted for each periodic check. When the data is graphed it does appear that there are two generations of moth per year. Moth numbers peak in the spring with a second, smaller peak in the autumn.

Monitoring results for the small vehicles group shows a drop in numbers each successive year except for the third year of monitoring, September–August 2008/9 (Fig 1). This small increase in moth catch is attributable to a change in pheromone product in January 2009, when a change was made to a more efficacious lure (Baxter 2008).

Data for the wooden vehicles group shows a steady drop in numbers year by year (Fig 2).

The large vehicle group caused concern as moth numbers increased each year for the first three years of monitoring (Fig 3). Investigation led to discovery of a large dust build-up underneath the immovable large transport items and within the ballast that surrounded them. As a result of a successful deep clean and maintenance programme, moth numbers dropped dramatically from 697 in the third year of monitoring down to 138 in the fourth and final year. Another reminder of the importance of good housekeeping measures in an integrated pest management programme.

Motorbikes and bicycles

To build on the success of the work in reducing the moth infestation within the vehicle collection, the fibre identification programme was extended to include motorbikes and bicycles. The older models frequently contained wool and horsehair within seat paddings. Survey of

Fig 3
Monitoring results for
webbing clothes moth in the
large vehicles group.

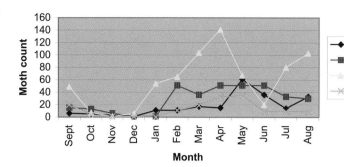

Large vehicles

Period	Total no. moth
Sept-Aug 2006/7	213
Sept-Aug 2007/8	339
Sept-Aug 2008/9	697
Sept-June 2009/10	146

Fig 3
Monitoring results for
webbing clothes moth in the
large vehicles group.

these collections discovered several active moth infestations within seats, and one within a sidecar. These infestations were treated by low-temperature pest control.

Gallery traps

In April 2009, to supplement the vehicle pest management work, webbing clothes moth pheromone traps were placed throughout the galleries in the old Museum of Transport. The aim was to adopt a mass-trapping approach. The old museum shared the Kelvinhall building with the International Athletics Arena, the two being separated only by an internal wall. Anecdotal evidence indicated that the moth infestation was also present in the athletics arena next door. It was hoped to prevent re-infestation of those vehicles that had been successfully treated by mass-trapping of the male moth population. Moth catch numbers were high in the beginning: April 2009 total of 427; May 2009 total of 418. The following year catch numbers had decreased: April 2010 total of 181; May 2010 total of 125. The reduction in numbers indicates a drop in the moth population, and some success with the mass-trapping approach.

IPM at the Riverside Museum

Pest management work at the old Museum of Transport was heavily focused on common webbing clothes moth due to the severe infestation. The move to the new Riverside Museum with its Clydeside location and improved environment opened up the possibility of infestations by other insect pests. Other Glasgow museum venues had experienced problems in the past with biscuit beetle (*Stegobium paniceum*) and carpet beetle (*Anthrenus verbasci*). The IPM strategy for the new museum needed to be extended to take account of possible infestations by all insect pests.

Risk rating Story displays

Within the IPM strategy for the Riverside Museum the vulnerability of the transport collection to infestation by clothes moth was extended to take account of the vulnerability of the whole collection on display to infestation by all insect pests. IPM work carried out at other museums (Doyle *et al* 2007; Pinniger 2007) applied risk zones to areas according to their vulnerability to infestation (either through the use of, or activities carried out within, that area; or due to the display or storage of artefacts depending on their construction materials). The design of the Riverside Museum meant that zoning was not possible as, internally, the building is a single undivided space.

Story displays were assessed according to the materials of the objects they contained and assigned a risk rating. Vehicles were treated as an individual display. Objects containing wool, fur, horsehair, feathers, skin, dried plants, cereals or legumes were given a high vulnerability rating; those with wood, card or paper were given a rating of medium vulnerability; synthetic and vegetable fibre materials were classified as low vulnerability; inorganic materials (glass, metal, stone) and rubbers or plastics were deemed to have no vulnerability. All displays and vehicles were classified as high, medium, or low vulnerability; then further sub-divided according to being on open or enclosed (cased, for example) display. This gave a total of six ratings, for example high vulnerability open display (or open vehicle); high vulnerability cased display (or enclosed vehicle); medium vulnerability open display, and so on.

Assignment of risk ratings influences the placement of pest traps, use of pheromone lures, periodic surveys and cleaning schedules. Highly vulnerable displays will be monitored more frequently than those of medium vulnerability. Displays of low and no vulnerability materials will be less frequently monitored, allowing resources to be concentrated where needed. Open displays will be visually checked and require regular maintenance to prevent dust build-up. Dust build-up, even on non-vulnerable materials can easily support clothes moth infestation, as has been seen in the old museum.

Challenges of monitoring and maintenance

The pest management work has massively reduced the severity and number of infestations within vehicles. Active vehicle infestations have more than halved; the small vehicles group has dropped from 25 vehicles infested down to 10. However, the vehicle collection still requires to be monitored and treatments carried out as necessary.

The innovative display of vehicles at height presents challenges for pest management. These will require regular maintenance to prevent dust build-up. The Car Wall, Velodrome, Motorbike Wall, and the Rest-and-Be-Thankful ramped car display will be accessed every three months to monitor for pest infestation. Equipment purchased for installation of the vehicles, a combilift four-way side-loader, will be used for maintenance and pest monitoring.

In addition to the story displays, areas of concern vulnerable to supporting pest infestations have been highlighted for maintenance and periodic survey. These include dead spaces under display furniture, the top of historic shop units, prop horses with real horsehair, and service areas containing electrical and other equipment. Also of concern is the acoustic lining of the building, a perforated sheet material that extends up to seven metres high on the internal wall. Maintenance will be required to keep the perforations dust-free.

Conclusion

The challenges for pest management presented by the new Riverside Museum in Glasgow highlights the importance of considering these issues at the planning and design stage of a major project. Alternatively, achievable maintenance programmes need to be put in place to prevent the development of pest infestations. Ease of access to the story displays greatly aids pest management and general maintenance of the collection and the building.

The importance of good housekeeping regimes for successful pest management cannot be stressed too strongly. Housekeeping of collection, non-collection and service areas should all be included within an integrated pest management strategy to combat and so keep pest infestations at bay.

Acknowledgements

The author would like to acknowledge the support of David Pinniger, consultant entomologist, and the IPM team for the Riverside Project – Rebecca Jackson-Hunt, Margaret Dobbie, Andrew Howe, and Jacek Wiklo.

References

Baxter, P, Imperial War Museum, pers comm, December 2008

Doyle, A M, Pinniger, D and Ryder, S 2007 'Risk zones for IPM: From concept to implementation'. *Collection Forum* **22**(1–2), 23–31

Pinniger, D 2007 'Using risk zones to develop IPM'. Presentation at *ICON Care of Collections Group workshop Pest Management in Practice,* Imperial War Museum, London, 25 October 2007

14

The problems of house mice in historic houses and museums

Ed Allan

Ed Allan Consultancy Services, Folkestone, UK, ed@edallan.co.uk

ABSTRACT

House mice (*Mus domesticus*) are one of a small number of commensal rodents in the UK and they have evolved alongside man and are almost always found living in close association with him. They are extremely adaptable, being able to live in deep freezes as well as hot, humid areas. Museums and historic buildings are very attractive to these rodents and they can cause serious and expensive problems when in the wrong places. There can be direct gnawing damage to structures, electrical cables and objects. Their urine and droppings can contaminate and stain materials and they can damage insulation and soft furnishings in their hunt for nesting materials. In catering areas they can spread food poisoning bacteria which could lead to sickness, fines, prosecution and even closure. Control of mice infestations can be very difficult to achieve. The use of conventional cereal-based rodenticides can bring new problems such as insect infestations. Dead mouse bodies can also support other insect species. Local infestations may be solved by trapping and removal, but the emphasis in control programmes must be manipulating conditions to discourage house mice. Such measures include good food and waste management, proofing to high standards, training staff to recognise the signs of house mouse activity and the need for good housekeeping.

KEYWORDS

House mice, museums, *Mus domesticus*, rodent, pest, monitoring, trapping

Introduction

The house mouse belongs to a group of mammals known as rodents. Of all the mammals, worldwide, rodents form over 40 per cent of the total number of species, and given that many species live in large colonies the actual numbers of rodents probably exceed any other group of mammals on Earth. There are estimated to be over 2,220 species of rodents (Wilson and Reader 2005), ranging in size from the African pigmy mouse weighing seven grams to the capybara weighing over 80kg. Rodents have two incisors in both their upper and lower jaw which grow continuously and must be kept short and sharp by constant gnawing; this is the origin of the name, from the Latin rodere, to gnaw. Rodents are significant in many ecosystems because they reproduce rapidly and can function as food sources for predators, as a mechanism for seed dispersal and as disease vectors. Humans use rodents as a source of fur as pets, as model organisms in animal testing, for food, and even for detecting landmines.

A small number of rodent species have adapted to living in close association with humans (commensal). When man began to grow and store crops they quickly exploited this source of food. Man responded by storing crops inside buildings, thus allowing rodents to move indoors and become pests.

In the UK there are three rodent species that commonly cause problems. Two of them are rats and the third is the house mouse. Whilst the brown or common rat (*Rattus norvegicus*) and the black or ship rat *(Rattus rattus)* could be a problem in museums and historic houses, they are more likely to be found outdoors exploiting poor waste management practices or over-zealous bird feeding. However, they will come into a building to search for food, though they normally leave unless they find an undisturbed area such as a roof void or under-floor cavity. Grey squirrels *(Sciurus carolinensis)* are often seen to display similar behaviour.

As the common house mouse is small and thrives inside buildings (as, occasionally, does the wood mouse *Apodemus sylvaticus*) it is most likely to live its entire life indoors where it will cause the most problems.

The house mouse

The house mouse is thought to have originated in the steppes of central Asia on the Iranian–USSR border area (Schwarz and Schwarz 1943) and from here it spread worldwide. In Europe there are two sub-species recognised – *Mus musculus domesticus* the western European house mouse, usually found in the UK, and *Mus musculus musculus* the eastern European house mouse. In the UK the house mouse has been identified from pre-Roman and Iron Age deposits and is believed to have arrived in Britain around 1000 BC (Meyer 2007).

This species is now common in a wide range of urban and rural buildings all over Britain. It is mainly a house-dweller and rarely lives outdoors or in sewers. It prefers an internal habitat as it does not compete well with small external mammals such as wood mice.

House mice vary in colour from grey/dark grey to almost black, the fur is soft and their tail is about the same length as the head and the body. The under-belly may be slightly paler than the fur above but is not usually markedly so.

House mice avoid wet or damp conditions, preferring dry environments because with their small size it is easier for them to keep warm and retain body heat. They are able to cope with cold conditions if the environment is dry and they have access to nesting material to make warm nests. Mice have been found living in deep freezes, where they develop longer and thicker fur to keep warm and feed selectively off high-energy foods. This adaptability to their surroundings is one reason for their worldwide success.

Family groups of house mice are dominated by a single male, which will attempt to exclude all other males that try to mate with his females. This dominant male will retain as many females within his family group as he can and will allow sub-dominant males to remain in the group as long as they do not challenge him by trying to mate with the females. As population density increases, aggressive behaviour tends to increase, both within and between family groups, and territory boundaries are more rigorously defended. Under more crowded conditions low-ranking or subordinate mice are less able to secure territories and will not breed. The communication of this dominance structure within a mouse colony is achieved by urine marking that enables mice to live in family-based territories. Therefore, colonies of mice in a locality become separated spatially due to their territorial behaviour.

Breeding in all rodents can be very rapid and the house mouse is no exception. Females are sexually mature at 8–12 weeks old and can give birth around every 21 days. With litters containing five to six young on average and six to eight litters per female, the populations can expand rapidly over a short time.

Potential problems

There are several ways in which house mice can cause problems in museums and historic buildings. Some are obvious such as direct damage to artefacts, and some are less obvious such as causing the closure of a public restaurant. Gnawing damage can be severe and permanent. Their teeth are very strong and the constant gnawing keeps them sharp. They can damage all types of wood as well as soft metals such as aluminium and lead. Mice do not taste anything that they gnaw. When breeding, they will shred soft materials to make nests for their young. Favourite nesting materials include shredded cardboard, paper, textiles and carpets, and loft insulation. In the loft and under floorboards mice can gnaw electrical cables leading to local disruption of electricity supplies or even fire, whilst the gnawing of water pipes can give rise to leakage leading to water damage.

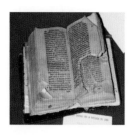

Fig 1
Mouse damage to old book.
(© David Pinniger)

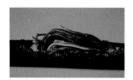

Fig 2
Gnawing damage to
electrical cable.
(© Ed Allan)

When mice are present in an area they produce fresh droppings and urine, which can contaminate and stain surfaces. During its lifetime, a house mouse will excrete about 0.5 litres of urine (Pinniger 2001). In addition, many diseases can be spread in this way, particularly if staff or visitors unwittingly come into contact with rodents or their by-products (Figs 1 and 2).

Most museums and historic houses have catering facilities that form an important revenue source as well as providing a service to visitors. Mice carry and spread food poisoning bacteria and so are potential causes of sickness in staff or visitors. The Food Safety Act 1990 and its associated Regulations are aimed at ensuring that food does not pose any threat to human health. The presence of house mice in or around catering areas can give rise to customer complaints leading to possible prosecution and even closure of the facilities.

When it comes to eradicating house mice, it is important to remember that the control measures can also cause problems. Poisonous baits or detector blocks, which are generally based on cereals, are also very attractive to a variety of insect pests. The poison does not affect the insect, and it positively thrives in baits, particularly old baits that have been left behind by pest contractors. Infestations of grain weevil, saw-toothed grain beetle, spider beetle and various moth species have been found in loose grain baits and in wax block baits, whilst carpet beetle larvae *(Anthrenus sp.)* have been found living in the blocks. These insects were breeding in the baits and spreading into textiles and natural history specimens (Pinniger 2001). In addition, dead house mice, which are not collected and disposed of, will attract carrion insects such as blowflies and dermestid beetles. As the bodies dry out they can become an attractive food source for carpet beetle larvae (woolly-bears).

Control strategies

Prevention rather than control is always the best strategy to adopt. Most of the damage caused by house mice to objects in a collection will be irreversible, so damage avoidance must be the approach. This can be achieved in two ways. Firstly, by eliminating the conditions that attract house mice to sites and that support infestations once they are there. Secondly, by involving all staff in pest awareness programmes so that signs of the presence of house mice are detected early and dealt with swiftly before any significant damage occurs. It is common for properties to employ a pest contractor for routine monitoring and control measures. However it is unlikely that contractor visits will be more frequent than monthly, allowing time for new infestations to establish and damage to occur. Staff can provide valuable 'extra eyes' to detect early signs of activity. To be effective, it is essential that staff have a simple pest reporting system so that sightings can be recorded and appropriate action taken.

Removing the cause

The conditions that house mice are seeking in order to successfully live and breed are the same for any pest. This consists of a suitable safe harbourage that is free of predators, and a suitable food and water supply (though house mice can survive on very little water). It is not difficult to see why museums and historic houses provide these resources in abundance. Whilst it is not practical to install predators to deal with house mouse infestations (cats would cause their own problems) the removal of food sources and the denial of access to suitable harbourage can be achieved. Good management practices around food areas are essential, particularly so when dealing with waste storage and removal. Food should not be left out, such as on open shelving, overnight, but must always be kept in mouse-proof containers. It should also be noted that small crumbs and morsels of food can support significant numbers of house mice and so cleaning floors and surfaces at the end of the day is much better than cleaning in the morning. Waste bins should be emptied or securely covered, and no dirty dishes should be left out overnight.

External areas are more difficult to manage, though house mice seldom go outdoors. However, poor waste management can encourage other pests such as rats, squirrels and birds. Where the public are allowed to picnic outdoors, then suitably large bins should be provided for waste food and packaging, and these should be emptied regularly and always before they are overfull. The feeding of birds around ponds and lakes should be discouraged as this will also attract rodents and insect pests.

The denial of access to suitable harbourage can be difficult as an adult mouse can squeeze through a gap of 12mm, and a young house mouse a gap as little as 6mm. There may be further difficulties because some of the most effective methods may be visually unacceptable in a historic building. Small holes in walls and floors should be patched or repaired with an appropriate filler, and larger holes can be filled with metal mesh or brickwork before being set in mortar. Pipes and cables should be tightly fitted when passing through walls and floors, and doors should be fit tightly in the frame. Bristle strips at the bottom of doors can deter house mice but are not a complete barrier. Airbricks set into cavity walls can be covered in 6mm metal mesh without reducing their ventilation effect.

Staff training

Training staff to recognise signs of house mouse activity is the most effective way to detect their presence early on and so initiate control measures before any serious damage occurs. Like most rodents, house mice are nocturnal and so are rarely seen during the daylight hours. There are many indicators which can tell you that mice are present without actually seeing them and a trained group of vigilant staff will be invaluable. Cleaning staff, in particular, can be useful as they will often be working in quiet, undisturbed areas where house mice will prefer to live.

House mouse droppings are one of the most obvious signs of their presence in a building. The spindle-shaped droppings are usually 3–7mm in length. They are commonly found on top of kitchen units and electrical panels, so demonstrating the good climbing skills of the house mouse. Fresh droppings are soft and shiny but within a few days they will become hard and dull. In roof spaces they could be confused with bat droppings, but bat droppings will be dry, crumbly and contain insect remains.

Footprints may also be seen in dusty areas such as under fitted units or on shelves and ledges. Dark smear marks may be visible in areas most used by house mice as they habitually use the same routes and leave greasy smears on the surfaces they regularly touch. These are most likely to be found on beams, around holes, on corners and at the base of doors and windows. House mice urine also has a very strong and distinctive smell, and this can be detected particularly on entering a closed room.

Staff should also look out for fresh signs of damaged and chewed materials that may be used for nesting. Rooms that are rarely visited should be included in regular inspections. Stored materials such as packaging and old display panels stored neatly so that any fresh signs can be easily spotted.

Control options

As previously mentioned, the long-term use of poison baits based on cereals is not recommended. Not only is there the risk of providing additional food and harbourage for a wide range of insect pests, but poisoned mice tend to die in inaccessible areas which can give rise to other insect pests and decomposition smells. In the UK's larger cities, house mice have developed an inability to digest their normally preferred cereal-based food and instead have shown a preference for high-protein foods, with the result that conventional baits may not work. They also show physiological resistance to some of the commonly used rodenticides, thus reducing the number of effective poisons. The use of rodenticides should only be carried out by trained persons and, ideally, be carried out by a pest contractor.

Pest contractors

When taking on a new pest contractor, it is essential that the outgoing contractor removes all their baits. Any missed should be removed and disposed of by the new contractor. The new contractor should be made aware of the special environments that exist in museums and historic buildings. Rodenticide baits should only be used for eradicating a persistent or new infestation and should be closely monitored and any bodies found removed and disposed of safely. A plan of the position of the bait points should always be drawn up so that missing ones are quickly spotted.

Monitoring mouse activity by using traps is a more practical approach for museums and historic buildings. With trapping, the animal is caught and therefore is not going to die elsewhere in a possible sensitive area. Traps can either be live capture or break-back type. Break-back traps are the most effective when used correctly. Traps can be hidden inside boxes, so staff and the public do not see them, and are set with the treadle-end placed against a wall or solid barrier along which the mice are moving. Food bait may be used but is not essential. Some mice become trap-shy if they are hurt or frightened by the noise of a springing trap, and so live capture traps may then work better. All traps need to be checked daily and mice removed and traps reset.

There are also traps known as sticky boards, which are considered inhumane by some people but can be useful in a particular tricky situation. Small boards pre-coated in glue are placed in areas of mouse activity and mice get stuck to the surface when they can be quickly killed. The British Pest Control Association (BPCA) has issued guidelines on the safe, effective and responsible use of sticky boards and it is essential that these guidelines are followed when this method is used so there is no undue suffering to the trapped animal.

Conclusion

The presence of house mice in museums and historic buildings has the potential to cause serious damage to both the contents of the building and the building itself. In rare circumstances the building could burn down. It is essential that there is a proactive pest control system in place that is aimed at gradually reducing the attractiveness of buildings to house mice and detecting mouse activity in its early stages before any significant damage occurs. Relying solely on a pest contractor is unlikely to prevent infestations occurring and it is essential to involve staff working in these premises and to have an effective reporting system to deal with any problems immediately.

Acknowledgements

My thanks to David Pinniger for the photograph of the mouse damage to the old book.

References

Meyer, A N 2007 'Commensal rodents'. *British Pest Management Manual* 1–88. British Pest Management Manual

Pinniger, D B 2001 *Pest Management in Museums, Archives and Historic Houses.* London: Archetype Publications, 83–100

Schwarz, E and Schwarz, H K 1943 'The wild and commensal stocks of the house mouse, *Mus musculus*'. *Linnaeus. Journal of Mammalogy* **24**, 59–72

Wilson, D E and Reeder, D M (eds) 2005 *Mammal Species of the World: A Taxonomic and Geographic Reference.* Baltimore: Johns Hopkins University Press

15

Minus 20 degrees in the sun

Anne Bancroft ACR*, Valerie Blyth ACR†
and Elizabeth F. Watson‡

V&A Museum, London, UK, a.bancroft@vam.ac.uk (author for correspondence)
†V&A Museum, London, UK, valblyth@vam.ac.uk
‡University of the West Indies, Barbados elizabeth.watson@cavehill.uwi.edu

ABSTRACT

This paper describes the low temperature treatment of a sizable insect-infested book collection in a tropical climate, using limited resources. Adapting and introducing best practice approaches of temperate climates to tropical institutions was a fundamental aspect of the planning. Researching realistic and cost-effective treatment options was a priority, as was finding appropriate expertise. Introducing and raising the profile of alternatives to methyl bromide fumigation was integral to the project, and strategies were put in place for the long-term care of the collection. Bulk low temperature treatment of cloth bindings proved to be effective, and key achievements within the success of the project include sustainability and transferral of skills.

KEYWORDS

Low temperature treatment, tropical collections, books, insect infestation, integrated pest management, collection migration

Historical background

The Richard B. Moore Collection located in Barbados represents some of the publications and other items that were owned by writer and bibliophile Richard B. Moore. Moore was a Barbadian, a communist, and the owner of a bookshop in Harlem, New York, which specialised in African and Afro-American titles (Wikipedia 2011). He was a prominent activist for African-American rights in New York from the 1920s. Moore helped to shape the New Negro Movement and the Harlem Renaissance, and was the author of texts on the Black experience. The collection was purchased by the philanthropic community group, The Lions Clubs of Barbados and given to the island as an independence gift in 1966. As a consequence, the University of the West Indies (UWI), the current caretakers of this collection, decided to submit a dossier on the collection to Barbados' National Register of UNESCO's Memory of the World Programme (MOWP) (Watson 2009). This Programme aims 'to guard against collective amnesia calling upon the preservation of the valuable archive holdings and library collections all over the world ensuring their wide dissemination' (UNESCO 2011). The programme's main objectives are 'to facilitate preservation by the most appropriate techniques; to assist universal access to documentary heritage and to increase awareness worldwide of the existence and significance of documentary heritage' (UNESCO 2011). The collection was housed in several locations, a number of which were not ideal for the items in the collection. Prior to returning to the UWI, the collection was 'temporarily' housed on the premises of the Barbados National Archives, where it remained for several years.

Climate and housing

In Barbados, the average temperature is 26.0°C with an average temperature fluctuation of ±2.5°C. The average humidity is 76 per cent and ranges from 71 per cent in February and May to 81 per cent in October and November. Prior to being treated, the collection was housed in a coral stone building with metal louvred windows. Two of the interior walls were wooden and the books were stored on wooden shelves. There was a defunct air-conditioning unit and evidence of past water damage on a few of the ceiling tiles.

The team

The chairman of the MOWP, Elizabeth Watson, was the project manager. While one author supervised the project on Barbados, the other acted as consultant from London. The local team consisted of Stanley Griffin, Assistant Archivist and Waveney McConney, Library Clerk. They undertook the management of the project after the low temperature treatment was completed.

Surveying the collection

Prior to conducting any preservation work on the collection, a sample survey was carried out. It determined the overall condition and the extent of work required to make the collection accessible. During the survey an active insect infestation was discovered. The collection consists of approximately 3,207 books, mainly cloth-bound. Approximately 68 per cent of the collection was affected by insect activity to varying degrees. Some bindings had neem leaves (*Azadirachta indica*) placed in the text blocks as a preventive measure against insect activity. The condition ratings of the books varied from 'fair' to 'very poor' (Webber and Norton 1998). Most of the damage to the books was found to be from insect activity and not from handling. In extreme cases the insects had burrowed through adjacent bindings, resulting in some book covers adhering to one another. The Barbados National Archives were informed of the infestation as soon as it was discovered. It was explained that the infestation could spread to their collection (Fig 1).

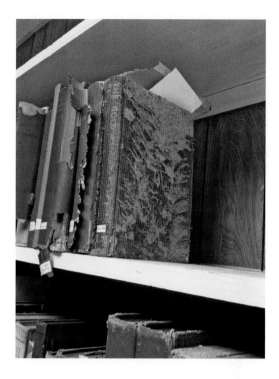

The severity of the damage required timely action to halt the infestation and prevent further damage. The MOWP was advised that immediate treatment was necessary and to contact a local entomologist to identify the active insects. Observations of the damage to the books revealed that silverfish were active. Additional insects were also suspected because of the extent of the damage and the presence of exit holes. A substantial amount of rodent excrement was also noted. In consultation with colleagues and experts in the field, coupled with personal experience and local knowledge, it was clear that it would be necessary to adapt current best practice approaches of treating insect infestation. Concerns were the amount of resources required, the shortage of expertise needed to deal with the infestation and whether it was possible to effectively treat the collection using current acceptable practices.

Locally available insect infestation treatments

Local treatment options for insect infestations are based on those used in the agricultural industry. The options are limited and reliant on fumigation. The local entomologist investigated the infestation but could not identify the insects and he recommended fumigation with methyl bromide as its effectiveness was proven. Such an approach is typical with this type of infestation. Despite the effectiveness of this method there are a number of negative issues associated with its use. There is a general lack of awareness regarding the health and safety risks associated with this chemical, as well as it being an ozone depleter. Despite these hazards, there appears to be a lack of support to ban it nationally or regionally and there was no knowledge of non-chemical alternatives. The preferred chemicals for this type of infestation in the local commercial sector would be sulfuryl fluoride and aluminium phosphide (Parsons 2009). The practice of non-chemical preventive measures, such as good housekeeping and the use of neem (Prasad 1986) and bay leaves (*Laurus nobilis, Lauraceae*) are now generally only used in domestic households. The fact that such chemicals were still being considered to treat infestations made it clear that this project could be the platform needed to showcase the effectiveness of local non-chemical treatments.

Non-chemical alternatives to treatment investigated

The treatment options were reviewed and a proposal and budget were drawn up. Wherever possible, expertise, materials and equipment were sourced locally. The treatment options were investigated by the team members in the UK. Current treatments commonly used in the UK would be effective but there was limited local expertise in these areas and no dedicated facilities or infrastructure available to treat the collection routinely. Heat treatments and anoxic environments are not available on Barbados. The possibility of importing a Controlled Atmosphere Technology (CAT) enclosure bubble from Rentokil® was ruled out due to expense. Vacuum-packing the books was also considered as a method of creating an anoxic environment for the books but this proved not to be a viable or practical solution. Unfortunately, there was no time to test all of the options available and low temperature, a proven and cost-effective treatment for insect infestations (Ackery 2000; CCI 1997), rapidly became the choice for treatment on the island.

Low temperature treatment facility

Contact was made with the island's cold storage company. It was explained that these objects were of national as well as international importance and that with their cooperation they would help to preserve their nation's cultural heritage. They were keen to be part of the project and gave us complimentary use of their facility.

The freezer units are purpose-built for large palletised perishable food products – mainly ice cream – but also anything from fish to Christmas trees. The building is a large warehouse divided into large rooms. The rooms are sealed by insulated and fireproof doors that can be secured. The level of security is very high, with restricted access to keys and monitoring by customs officials. Only authorised personnel are allowed into the freezers and they are required to wear personal protective equipment. A request was made for two members of the project team to have access to the units. This was to assess whether this storage environment would put the books at risk, to see how the books should be packed, to locate the best area for them and confirm the unit had the consistent temperature of –25°C that was stated (Fig 2).

Approach

The approach of the project was based on the initial survey report and images from it. In order to treat the infestation successfully it was necessary to know what the insects were. The local entomologist was unable to positively identify the active insects. Time and geographical constraints meant the treatment would have to go ahead without identification. Low temperature treatment of library material has been established for many years with many documented examples of success. The team therefore felt confident that it would work with this collection. After the collection was treated it would be moved directly to improved housing. Because the collection was not returning to the infested site, it meant there was less immediate pressure to determine the source of the infestation and to treat the site. The size and layout of the room restricted the number of people who could work on the project. The responsibility and introduction of risk

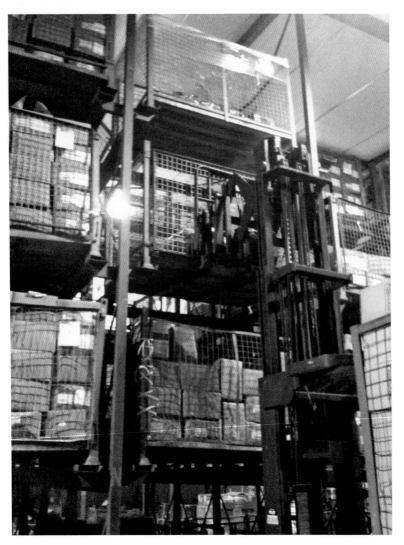

Fig 2
Interior of the cold storage unit.
(Photo: Anne Bancroft)

still necessary to have their expertise so that the project would be monitored and reassessed if necessary throughout the treatment. It was decided to keep in contact with the preventive conservator using all forms of current communications systems available. As and when needed, the authors could confer on Skype®, use digital images, videos and live electronic streaming.

Training

The team was given an overview of the project and the collection, along with practical training together with supporting literature. An introduction to integrated pest management (IPM) with an explanation of the various non-chemical treatment options for infestations was provided, supported by demonstrations on how to prepare the books for the treatment. Training in book handling, reduction of surface dirt, recognition of damage, measuring for phase boxes and collection migration was also included.

Treatment options decisions and proposal

Blunder traps were installed before work started so the active insects could be retrospectively identified. An assessment of the new storage space was made and readings of the temperature and relative humidity were collected and analyzed. Various options to prepare the books prior to the low temperature treatment were investigated. Initially, it was decided that the books should be cleaned before treatment to prevent the possible risk of the dust becoming engrained on the surfaces. However, after consultation with another expert in the field (Florian 2009) it was realised that doing so would run the risk of the insects seeking sanctuary in the archives collections stored on the same premises. As the number of books affected was large, palletisation of the books on site was investigated. It would have involved wrapping them in acid-free paper and polyethylene directly on pallets that could be transported to the cold storage facility. This, too, was ruled out because of the cost of renting a forklift and driver. The most appropriate option was to prepare them in bulk packages

assessments lay with the lead conservators and it was decided that, wherever possible, the standards should follow UK best practice.

A team of six people was needed to prepare the books for treatment but there was no dedicated staff to draw on. Personnel from local institutions such as the UWI, the Barbados Museum (BMHS), the National Art Gallery, the National Archives and the National Library Service were informed about the proposed treatment and invited to be part of the team. The treatment would also act as a practical workshop to provide local expertise and aid with professional development. Elizabeth Watson secured future housing for the collection at UWI, which meant their staff would have more access to the project and provide extra resources.

Due to financial constraints, both lead conservators could not be present. However, it was

that could be lifted manually. These were then transported to the cold storage company and palletised there, supervised by a conservator and a team member.

Materials and equipment

Sourcing and importing materials and equipment was challenging at times. Most materials and some equipment needed to be imported. It was not possible for the authors to physically check the materials that were being sourced locally. However, this was done effectively by the locally based team leader after consultation with the international experts.

Time estimate and treatment proposal

The project was estimated at nine days in total. This was broken down as one day for training, three days packing and one day for writing the report. The books would be treated for three days at –25°C as that was the temperature in the cold storage facility. The books would be kept in the polyethylene packages for one day while they returned to room temperature. An additional day would be allowed for unpacking.

After treatment the books would be cleaned, checked for insect activity, condition assessed and given recorded condition ratings to identify conservation priorities. The books with structural damage would be tied using linen tape and measured for boxing. The space was monitored for insect activity. It has been proposed to re-catalogue the collection electronically and to source a budget for conservation.

Treatment record

Owing to the unforeseen circumstance of an additional collection being housed in the same rooms as the Moore collection, it was necessary to revisit the time estimate. The custodians of the other collection blocked access to the room, damaged some of the books in the Moore collection and had to be informed that they were in danger of infesting their own collection. This situation meant that there was even

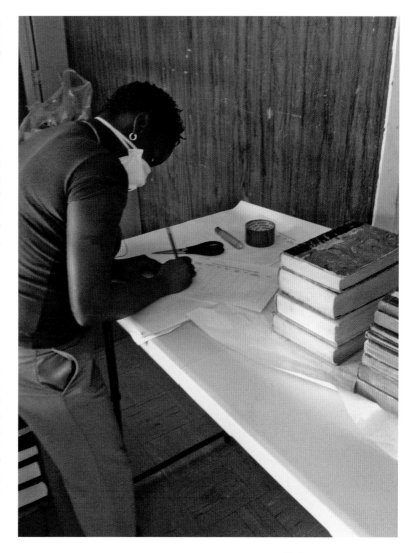

Fig 3
Team member preparing books for low temperature treatment.
(Photo: Anne Bancroft)

less working space. It was not possible to work as many hours per day as initially envisaged because of the levels of heat and dust. A team of eight people were trained to prepare the books for low temperature treatment, normally working in pairs. The number of people working at any given time varied and consequently the productivity varied (Fig 3).

The most efficient method was to assign a bay of shelves to each pair. Books were moved firstly from the bottom shelf, because if started from the top, the dust and frass would fall on the books below so increasing the need for further cleaning. They were wrapped in packages using acid-free paper and polyethylene sheeting, with approximately 10 books per package. The number of books wrapped in each parcel was recorded. It was decided not to record the pressmarks of the books as this would slow the progress. Any visible insects/

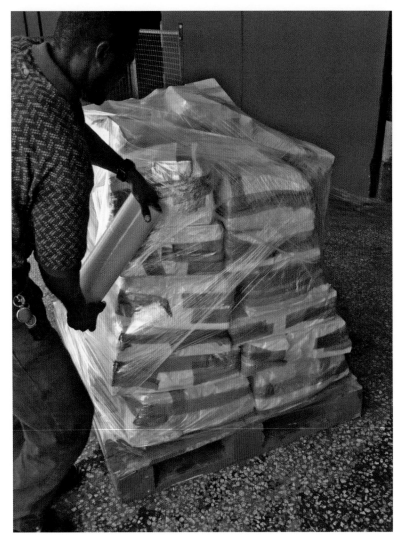

Fig 4
Palletisation of the wrapped books.
(Photo: Anne Bancroft)

larvae were collected where possible. Each package was sealed with brown pressure-sensitive tape and labelled with the date and a running package number. A decision was made to place all of the books that were wrapped at the end of each day into the cold store immediately because of concern that the condensation could develop on the inside of the package and there was a risk of mould growth (Fig 4).

Transportation and the cold storage unit

Team members used their own cars to transport the books as there was not a suitable enclosed van available. The packages were placed in plastic storage boxes and transported to the cold storage unit daily. At the unit, the boxes were stacked on a pallet and wrapped in cellophane to ensure minimum movement when the pallet was transported by forklift truck. A member of the conservation team escorted the palletised books into their dedicated holding pen and kept in cold storage. All of the books remained in cold storage until the last pallet of books was exposed to –25°C for three days. The books on the pallets were all removed from cold store at the same time.

After low temperature treatment

The books were transported to the library of the UWI. Upon reaching ambient temperature the books were unwrapped and cleaned using soft natural-hair bristle brushes. The removed dirt and insect frass was checked for insects and then vacuumed. There were not enough team members available to unwrap the packages, which caused concern as the longer the books remained wrapped the greater the risk of fungal attack.

Identification of insects was carried out in London by Val Blyth and corroborated by David Pinniger, Consultant Entomologist to the V&A museum. The active insects were mostly cigarette beetle *Lasioderma serricorne* and silverfish *Lepisma saccharina*.

After the successful treatment of this collection, support was rallied to run an IPM workshop in the future. This will be a practical workshop, held at the BMHS based on its collections, storage and exhibition space. Staff in other institutions and commercial pest control companies will be invited to attend.

Conclusion

Securing funding is always challenging and in the current financial climate resources will continue to be restricted. Working as a consultant when based in another country adds additional stresses due to differing culture and resources available. The UK conservation experience is based on European approaches, approaches which are based on temperate climates and proven through decades of professional work. There was no guarantee that these would translate or indeed be appropriate in a tropical climate. Local expertise and a bespoke approach were crucial. Both lead conservators on the project are accustomed to working with colleagues that have a solid understanding of preservation, conservation and IPM. In Barbados, it was necessary to introduce and make a case for non-chemical treatments. Securing materials and equipment was problematic, time-consuming and costly. What became clear was the need for adaptability and innovation. A consultant's involvement is limited and expensive, and one had to be very clear when specifying what materials and equipment were needed. More importantly, ensuring that the local team/custodians were imparted the skills they need to ensure the sustainability of the project was a fundamental outcome of the work.

Acknowledgements

The project's success is due to Elizabeth Watson's effective and dynamic management. We would also like to thank for their input Catt Baum, Victoria Button, Alison Callender, Alissandra Cummins, Mary-Lou Florian, Kerron Hamblin, Waveney McConney, Stanley Griffin, David Pinniger and Jane Rutherston.

References

Ackery, P 2000 'Are low temperature pest disinfestations procedures safe?' *The Biology Curator* **8**, 28–30

Canadian Conservation Institute (CCI) 1997 'Controlling Insect Pests with Low Temperature' CCI Notes 3/3. www.cci-icc. gc.ca/publications/ccinotes/enotes-pdf/3-3_e.pdf (accessed 4 October 2011)

Florian M-L E, pers comm, 2009

Parsons Pest Control, pers comm, 2009

Prasad, L K 1986 'Role of neem leaves in protecting textile material and paper documents'. *Conservation of Cultural Property in India* **1**, 14–15

United Nations Educational, Scientific and Cultural Organization, Memory of the World Programme 2011 http://portal. unesco.org/ci/en/ev.php-URL_ID=1538&URL_DO=DO_TOPIC&URL_SECTION=201.html (accessed 31 August 2011)

Watson, Elizabeth, UWI Campus Librarian, pers comm, 2009

Webber, P and Norton, A 1998 'The power of the poster and paper conservation'. *Conservation Journal* **29(3)**, 13–17

Wikipedia 2011 http://en.wikipedia.org/ wiki/Richard_B._Moore (accessed 31 August 2011

16

Mobigas at the National Gallery of Victoria, Australia and the struggle for recognition by quarantine authorities

Janelle Borig

National Gallery of Victoria, Melbourne, Australia. janelle.borig@ngv.vic.gov.au
www.ngv.vic.gov.au

ABSTRACT

This paper introduces the features of the National Gallery of Victoria's (NGV) anoxic pest treatment facility, Mobigas, that it purchased and imported from Germany in 2006. The NGV is pleased with the zippered tent chambers; the automated and mobile nature of the entire unit, and the processes implemented for its efficient operation. The only concerns are for future maintenance due to the problem of distance from the manufacturer. The paper then addresses the process the NGV has recently undergone in applying for acceptance as a Quarantine Approved Premise through the Australian Quarantine Inspection Service. Approval was granted; however, a secondary application to have Mobigas granted status as a quarantine approved treatment facility was declined on concerns relating to calibration and data logging. An application to attain 'blanket approval' of Mobigas to treat imported infested or high-risk artworks will be resubmitted in the near future.

KEYWORDS

Anoxia, quarantine, import, Mobigas, AQIS

The Mobigas anoxic pest treatment system

In 2006, the National Gallery of Victoria (NGV), in Melbourne, Australia, purchased Mobigas, an anoxic treatment system manufactured in Germany. Until this time, pest-infested or high-risk items were treated either in a small chest freezer or an in-house constructed anoxic system using temporary Cryovak bags. Larger items had to be sent off-site to contractors for pest treatment, usually using methyl bromide.

Given the handling, safety and insurance risks of off-site treatment and the increase in art acquisitions from northern Australia, the Torres Straits and surrounding tropical regions, the NGV considered it necessary to expand its pest-treatment capacity by finding a system which would safely treat large volumes of material of widely varying organic matter. Anoxia was the preferred methodology and various locally available systems were explored. The options available were large, rigid chambers, such as reconfigured refrigerated shipping containers, which proved to be extremely expensive and required large volumes of gas, and foil or film-based chambers, which were deemed too impractical and wasteful to construct for the volumes of artworks being dealt with.

The Mobigas system employs polyurethane, UV-sealed 'tents' with access through a gas-tight zip. The sizes of these chambers can be custom-made to clients' wishes and air can be evacuated via an in-built ball-valve to shrink the tent around the contents to a safe level, reducing the volume so there is less gas wastage. It was these features, along with, as the name suggests, the mobility of the equipment which persuaded the NGV to invest in this system.

The NGV first learned of Mobigas through the author, who had used them when employed at the Bayerisches National Museum, in Munich, Germany. This institution had developed the prototype system with Frank Paukstat, owner and founder of P+S Labormedien, and had used it successfully for nearly 10 years prior to the NGV's purchase.

The NGV's version consists of two aluminium cases, one containing the computerised control unit with digital display of internal conditions and in-built datalogger; the other containing the bubble-humidification unit. Cylinders of high-purity nitrogen mounted on a trolley provide the gas source. There are four ports built into a tent: one for gas entry, one for emergency pressure relief, one ball-valve port for air evacuation and one for internal sensor connection. There are two sensors, one measuring temperature and humidity, and the other is an electrochemical cell measuring oxygen levels.

Three tents of varying sizes were purchased by the NGV. The small tent: 700 × 2,200 × 1,100mm, was specifically made to accommodate costume boxes, stacked four high (Fig 1). The medium tent: 2,200 × 2,600 × 1,150mm, can house tall and bulky items, whereas the largest: 1,500 × 6,000 × 1,200mm, allows long works such as loosely rolled carpets, indigenous canoes and poles to be treated. An external metal framework is constructed from which the tent is suspended, aiding in the loading process. Packing is optimised to keep the volume to a minimum and internal structures are often utilised to ensure the safety of the objects. The tent is unhooked prior to operation and excess air is evacuated.

There is a continual process of evacuating the oxygen–nitrogen mix using a vacuum cleaner through the ball-valve port during the filling process, until the pre-set residual oxygen level is reached. The system then operates automatically, issuing more nitrogen as required to maintain the set levels. Staff, however, monitor levels daily; releasing gas out under a fume extraction unit when the tent starts puffing-up, and exchanging the cylinders as required.

The main offending pests are variegated and black carpet beetle, common webbing clothes moth and silverfish. NGV operates Mobigas in a room at a constant 21°C, with a humidity set-point of 50–55%RH and with an oxygen set-point of 0.3–0.5 per cent. Artworks are treated for a conservative eight weeks. Treatment effectiveness is determined only visually, however, there have been no active pest reoccurrences post-treatment since operations began.

The gallery database (Vernon Systems) has been set up to create a list where each artwork identified as requiring treatment is entered and a priority date given for when it is required for other purposes. Items are sealed and stored in a dedicated quarantine room while waiting to be treated. When enough works are entered into the database list and it is approximately 10 weeks out from an item being needed, a

Fig 1
Set up of Mobigas and the small tent.

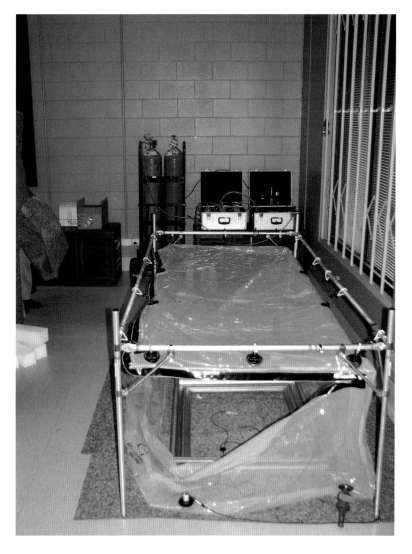

treatment run is arranged and the appropriate sized tent is selected. At the completion of each run, a treatment note is recorded against each artwork file, updating its history and enabling important treatment statistics to be easily retrieved.

The small-scale and customised nature of each Mobigas system meant that there was no standard operating manual. A manual was prepared in German by the manufacturer and subsequently translated into English by two German-speaking NGV staff members.

The author conducts regular training sessions and refresher courses for conservation and registration staff in the set-up and operation of Mobigas. Approximately ten conservators are now proficient in its use and three registrars have been trained to monitor and evacuate gases as required and assist in cylinder change. The conservation laboratory with the most items to be treated is responsible for supervising each treatment run.

The NGV has been very satisfied with the ease of operation and performance of Mobigas to date. The tents roll up and are stored inside plastic crates, the framework fits into tent pole sacks, the cases are reasonably lightweight and the trolley-mounted cylinders all allow for easy storage and transportation. The ability to treat infestations at various NGV sites as well as Victorian regional galleries was a valuable justification tool in the purchase proposal. Victorian state Occupational Health and Safety (OH&S) requirements, however, require detailed facility risk assessments prior to its use at any venue, placing restrictions on its versatility to respond quickly to off-site pest infestations. The State Library of Queensland (SLQ), Australia, purchased a similar version of Mobigas in 2009 after learning of the NGV's satisfaction with the system.

The NGV has been fortunate in that there have not been maintenance issues with the system to date. The oxygen sensor cells are replaced every two years, sourced from the original supplier in Germany, Krah & Grote. The NGV is extremely grateful for the email and telephone support given by the manufacturer, especially during the early stages of implementation. However, reliance on a manufacturer being a sole-operator and physically half a world away is a little unnerving. It will be interesting to see how any future system or mechanical faults are resolved.

Application to become a Quarantine Approved Premise

The Australian Quarantine Inspection Service (AQIS), along with Biosecurity Australia are agencies of the Australian Government's Department of Agriculture, Fisheries and Forestry (DAFF). The role of AQIS is to lessen the risk of foreign pests and diseases entering Australia by managing quarantine border controls. AQIS also provide inspection services to importers and exporters, and certification in order to retain the positive flora, fauna and human health status of Australia and broad access to export markets overseas (AQIS 2010).

Artworks that are purchased overseas or collection items that return from overseas loan are subject to import regulations and require quarantine clearance by AQIS. The NGV had a long-standing relationship with AQIS whereby an officer inspected incoming artworks that were flagged high-risk by AQIS at the NGV as it was being unpacked. However, due to the increase in occurrences, and to better facilitate the process, it was strongly recommended by AQIS that the NGV apply to become a Quarantine Approved Premise (QAP). As a QAP, many importing restrictions can be avoided and registered sites are trusted in providing an appropriate level of security and control against the introduction of exotic pests (AQIS 2010).

The application documentation certainly seemed targeted more for larger-scale importing firms and the criteria did not always seem relevant to NGV facilities. For example, the provision or access to wash-bays for containers. It was an arduous process taking nearly two years for the NGV to meet the requirements for QAP certification. It necessitated a certain infrastructure such as providing officer car parking bays and an inspection area with specified furniture, lighting and signage. Well-functioning integrated pest management (IPM) programmes had to be in place and internal procedures had to be developed to clearly demonstrate how each step of the qualifying criteria would be managed. Compliance included ensuring appropriate:

- isolation of quarantine goods and premise hygiene to restrict contamination
- movement of quarantine goods to and within a QAP and release from quarantine

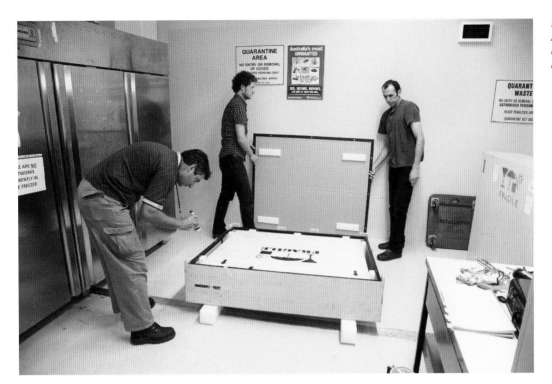

Fig 2
AQIS inspection in dedicated quarantine area at the NGV.

- waste management, ensuring the correct disposal of contaminated material

- accredited persons, two or more staff members trained by AQIS to directly supervise all activities involving the goods

- identification and awareness through correct signage, labelling and informing other staff

- traceability through sound record keeping

- contingency plan.

For the privilege of certification, QAPs are subject to annual and spontaneous audits, and must pay a yearly fee.

The NGV achieved Approval of Place for Quarantine in Class 1.3 Sea and Air Freight Depot Operations effective from 1 July 2010. Under this class of QAP, the NGV premises are only approved for its specific operations. There is now a dedicated quarantine area in the temporary storage depot. Signage and bright yellow paint clearly identify it and the area can be used for no other purpose than facilitating the inspection process. Eight registration and conservation staff members have completed the day-long AQIS training to implement their set requirements and to ensure someone is always available to receive shipments and prepare them for inspection. An AQIS officer still must come to the NGV and inspect each imported item and release it from quarantine (Fig 2). If the officer declares the artwork infested or high-risk, procedures are in place to activate the treatment process and eradicate the pest.

Standard AQIS approved methods of pest treatment are:

- methyl bromide fumigation

- heat sterilisation

- gamma radiation.

Application to have Mobigas deemed a quarantine approved treatment facility

Deterred by the unsuitability of the AQIS treatment methods for the NGV's varied collection, an application to have Mobigas recognised as a quarantine approved treatment facility was submitted in 2009. The NGV's internally prepared operating manual and procedure documents were supplied along with the application. A request for literary evidence of the effectiveness of anoxia and Mobigas was subsequently requested. A copy of the publication *Inert Gases in the Control of Museum Pests* (Selwitz *et al* 1998) was the primary source provided.

Unfortunately, after a wait of over a year, the NGV received an email (Orbell 2010) from AQIS stating:

Mobigas does not have blanket approval as an AQIS approved treatment … as there are issues with the use of this system, such as calibration and data logging … Mobigas will need to be approved on a case-by-case basis until further direction by AQIS … The QAP approval does not sanction general use of Mobigas treatment of quarantined goods at the gallery.

This disheartening news means that any time an AQIS officer suspects infestation, the artworks must be sent off-site for treatment or an application is submitted to use Mobigas on-site, monitored by an AQIS officer (with a charge for their time). Fortunately, the single occasion in which AQIS identified a threat and called for treatment, Mobigas was granted one-off approval for use and the outcome was deemed successful. It is assumed that anoxia methodology is given little attention by AQIS experts because of the long treatment duration in comparison to other methods, and so it is not considered a viable option for AQIS to use. The sensitive nature and small scale of art import does not seem to warrant detailed investigation.

The NGV continues to use Mobigas to treat pest threats to collection items which remain within Australian borders and is hopeful to reapply to AQIS for blanket approval in the future.

Conclusion

The NGV recently went through a process of readying its facilities and procedures in order to comply with the criteria of becoming a QAP: Class 1.3 Sea and Air Freight Depot Operations through AQIS. This was granted in July 2010. A secondary application to have the NGV's anoxic pest treatment facility, Mobigas, that was purchased from P+S Labormedien in Germany in 2006, accepted as an AQIS approved treatment facility was rejected on the grounds of concerns about calibration and data logging.

All flagged, imported artworks can be inspected on NGV premises. However, if an infestation is suspected by AQIS, it must be treated at an off-site facility using an AQIS approved method, usually methyl bromide fumigation. It is possible to apply for approval to use Mobigas on a case-by-case basis.

The NGV is thoroughly pleased with the performance of its automated and mobile anoxic system and the efficient processes it now has in place for its operation. The only concerns about the system are the reliance on the sole and distant manufacturer for ongoing support. The NGV will continue to use Mobigas to treat its non-imported collection items and intends to reapply to have Mobigas methodology, and indeed anoxia in general, reviewed and hopefully granted blanket approval in the near future.

References

Australian Government, Australian Quarantine and Inspection Service, 'About AQIS' (22 October 2010). www.daff.gov.au/aqis/about (accessed 24 February 2011)

Australian Government, Australian Quarantine and Inspection Service, 'Quarantine Approved Premises' (23 November 2010). www.daff.gov.au/aqis/import/general-info/qap (accessed 24 February 2011)

Orbell, G AQIS, 2010 Manager – Industry Partnership Unit, Victoria, pers comm, 12 August 2010

Selwitz, C and Maekawa, S 1998 *Inert Gases in the Control of Museum Pests*. Los Angeles: The Getty Conservation Institute www.getty.edu/conservation/publications/pdf_publications/inertgases.pdf (accessed 31 August)

17

Assessment of the Thermo-Lignum oven pest eradication treatment on natural and synthetic polymers and resins

Dr Marieanne Davy Ball ACR*, Christina Bisulca† and Dr Nancy Odegaard‡

*Kulturhistorisk Museum, Oslo, Norway, Marieanneball@yahoo.co.uk
(author for correspondence)
†Arizona State Museum, Tucson, AZ, USA, cbisulca@email.arizona.com
‡Arizona State Museum, Tucson, USA, odegaard@email.arizona.com

ABSTRACT

The Kulturhistorisk Museum (KHM) has recently acquired the Thermo-Lignum heat oven for pest eradication of collections. The heat treatment works by raising the temperature to above that at which insects can sustain life and at elevated relative humidity (RH) to prevent drying of materials during the treatment. While most objects appear to be stable using this treatment, conservators have noted changes that prompted more detailed investigation of the effect of this procedure. The preliminary investigation at KHM was an assessment of alteration to 34 resins, waxes and adhesive samples that are commonly used in ethnographic collections and in conservation treatments. These materials were treated with the same procedures used in the Thermo-Lignum process used for collections. Changes to samples were monitored by visual inspection, weight and dimensional changes, and chemical alteration assessed using Fourier Transform Infrared Spectroscopy (FTIR). These initial results indicate that the Thermo-Lignum procedure can have detrimental effects in proteinaceous adhesives. Acrylics were prone to slippage in joints, epoxies yellowed, and many natural resins and fats showed colour change after treatment. On actual collection materials, analysis of an efflorescence that formed on the surface of some wooden objects after treatment identified the migration of fats during heating. This information along with ongoing analyses at KHM can provide the necessary information on each material's response in order to make informed decisions about objects that can be safely treated.

KEYWORDS

Thermo-Lignum, adhesives, resins, proteins, deterioration, pest eradication

Background

In 2009 the Kulturhistorisk Museum (KHM) at the University of Oslo built a new storage facility for its ethnographic, archaeological and antiquities collections. Concurrently, the museum also upgraded its pest eradication systems to ensure that the objects are pest-free before relocation to the new permanent storage areas. After much debate and comparative analysis of available treatment methods, it was decided to use two different systems in tandem: a walk-in freezer and a heat chamber.

The premise of heat eradication is that an object is heated to a temperature above which typical museum insect pests cannot sustain life (52–54°C), while keeping the relative humidity (RH) constant to prevent excessive drying and distortion to the objects (Strang 1992). The heat chamber chosen was from Thermo Lignum, which has an internal capacity of 38 cubic metres allowing very large objects or multiple objects to be treated simultaneously. The RH can be set to any desired level, which at KHM is usually between 40 and 50%RH to correspond to storage conditions. The oven ramps up temperature gradually and holds at the maximum temperature for a set time (90 minutes in the KHM protocol). The primary advantage of using the Thermo-Lignum procedure is that it is highly effective in killing insects yet is more time efficient than freezing, as the entire process takes 16 hours as opposed to the week needed for freezing treatments.

Given the high susceptibility of pest infestation in ethnographic collections, these objects were the first to undergo the new pest eradication protocols. Over a two-year period, around 25,000 objects have undergone eradication, the largest proportion in the heat chamber. Initial assessment of heat treatment was made shortly after installation, and few visual changes were noted to objects after treatment. However, with the treatment of more objects, visual observations raised questions about the routine use of this method for pest eradication. Significant deformations occurred to shaped keratin artefacts, which caused concern for induced deterioration to proteinaceous materials. A white efflorescence (bloom) formed on the surface of some objects after heat treatment, particularly on wooden objects from Africa (Fig 1). Due to these experiences and others, several

classes of materials were placed on to a list of those not to be put through heat eradication pending further analysis.

Heat treatment protocols for pest eradication were developed for use in the food processing and lumber industries. Most of the available literature pertaining to the Thermo-Lignum system specifically for museum collections is focused on its effectiveness in insect eradication. Fewer studies have addressed the effect of heat pest eradication treatments on specific collection materials. Tscherne and Schachenhofer (2008) found minimal damage to painted and gilt surfaces, with the exception of shellac. Ackery et al (2004) demonstrated that Thermo-Lignum treatment did not adversely affect DNA in entomological collections, however, the potential for movement of greasy materials was observed. The findings of Hardboard (1999) showed deterioration on fish skin specimens from using heat, but not on skin/leather with shrinkage temperatures above 50°C.

In conservation it has long been known that heat and humidity are primary agents of deterioration (Feller 1967; Feller 1994). A full understanding of the effect of heat treatment requires a detailed knowledge of how each material's equilibrium moisture content (EMC) responds to changes in temperature and humidity. The size of the material and its thermal diffusivity are also important considerations in determining appropriate protocols for heat treatments (Wang 2010). However, this information is not readily available for most materials, particularly those encountered in museums that are already deteriorated from natural ageing. There are numerous accelerated ageing studies of different materials available in the literature, but the parameters used in specific studies do not often coincide with the duration, temperature and humidity used in the Thermo-Lignum treatment.

At KHM there are many concerns that have been realised in the use of heat treatment. These concerns are compounded by the fact that many materials in artefacts may be under additional stresses (tension, sheer, and so on), and may be already deteriorated from natural ageing. The latter is of particular concern, as KHM some artefacts were collected about 200 years ago, and many objects may have undergone historic treatment(s) or have already been exposed to high temperature or humidity prior to acquisition. Moreover in the KHM pest

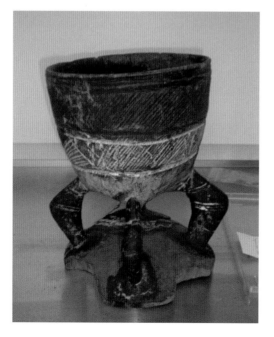

Fig 1
Bloom formation on wooden sculpture (UEM44130) after heat treatment.
(Courtesy of KHM: MDB)

management protocols, objects may go through several heat treatments as they are removed and/or returned for exhibition, study or loan. The primary concerns addressed for initial testing are:

- inducing or accelerating ageing processes in our collection. Many polymers are subject to chain-scissioning (depolymerisation) and oxidation during heat ageing, which can lead to a loss of mechanical strength. The deterioration (hydrolysis) and/or denaturation of proteins are of particular concern, as it is caused by heat and moisture ('gelatinisation').

- damage due to softening in cases where the material's glass transition temperature (Tg) is in the range of the Thermo-Lignum oven. The maximum temperature of treatment is above the Tg of many acrylics routinely used in conservation treatments, which can result in failure of joints or distortion of surface coatings. Of greater concern is that the Tg of polymers decreases with increasing RH and thus moisture content; a phenomenon that is most pronounced for hydrophilic polymers like proteins, certain epoxies, and carbohydrates. This has a dramatic effect in proteins. In the case of gelatine, the Tg decreases from over 200ºC at 0%RH to room temperature at 75%RH (McCormick-Goodhart 1996).

- alteration of water content. The EMC and sorption isotherms at elevated temperatures and humidity are not precisely known for many materials encountered in collections, particularly when considering many aged materials. The changes in water sorption with elevated temperature at 50%RH could result in dimensional changes (swelling/shrinkage) or a change in the moisture re-uptake of the material. Loss of water is a particular concern, and could result in embrittlement.

Methodology

Tests were carried out within the parameters used in the Thermo-Lignum heat chamber at KHM (core temperature of 54°C at 45%RH, 90 minute hold at maximum temperature).

Ambient temperature of the chamber was also measured using a maximum/minimum thermometer placed in the chamber. The smallest sized monitoring block was chosen due to sample size (refer to Thermo-Lignum® nd for further details on this process).

Preparation

The first tests were on 20 different adhesives (Table 1). Several preparations of each were prepared: (1) as an adhesive to adhere glass slides together, (2) as an adhesive for wood tongue depressors. In both cases, the adhesive was used to create a lap joint under light tension by suspending them in the chamber, allowing measurement of slippage. The use of

Table 1 Summary of results for Thermo-Lignum treatment on 20 adhesives

Sample	Adhesive	Glass % Inc. length (cm)	Filter paper % Inc. weight (g)	Colour change	Physical change	Change?
Acrylics	Paraloid B72	0.87				X
	Paraloid B44		0.11			X
	Paraloid B67	1.00				X
	Plextol B500					
Vinyls	Lineco PVA		−0.20			X
	Mowital B30H	0.48				
Epoxies	Araldite rapid			X	X	
(thermosets)	Akemi resin			X	X	
	Hxtal NYL – 1			X	X	
Proteins	Bone glue		−0.60	X	X	X
	Hide glue		−0.40	X	X	X
	Rabbitskin glue	0.86	−0.40	X	X	X
	Sturgeon glue	0.90	−0.20	X	X	
	Fish glue	0.45	0.10			X
	Gelatine	0.43				X
Cellulostics	Cellulose nitrate	0.83				X
	CMC					
	Klucel G					
Other	Gum arabic		−0.41			X
	Rice starch glue		−0.50			X

Table 2 Summary of results for Thermo-Lignum treatment on 14 resins and waxes

		Weight change	Colour/ surface changes			Indication of melt			
	Adhesive		Glass	Wood	Paper	Glass	Wood	Paper	Change?
Resins	Dammar	X							X
	Gum elmi								
	Shellac				X	X			X
	Copal								
	Lacquer	X	X						X
	Colophony		X	X	X				X
Waxes	Benzoin					X			X
	Beeswax	X					X	X	X
	Paraffin wax			X		X	X	X	X
	Microcrystaline wax						X	X	X
	Renaissance wax								
	Cosmolloid H 80						X	X	X
Fat/oil	Shea butter					X	X	X	X
	Cocoa butter					X	X	X	X

different substrates allowed for comparison between absorbent and non-absorbent materials in adhesion. (3) Samples absorbed into the centre of a piece of filter paper, the edge demarked in pencil, to monitor spread, colour or dimensional change. (4) Samples prepared as films on silicon release paper, then placed in glass vials. Fourteen resins and waxes were sampled (Table 2). Each material was painted on to a glass slide and a wooden tongue depressor to see if the substrate affected the result. A sample was spread on to filter paper to observe colour change and spread; and samples of film made for later FTIR analysis. In all cases a sample was removed as a control for FTIR analysis. The samples were allowed to dry for more than a week before being photographed against a colour chart, measured, and weighed.

Test procedures

All samples were subject to five cycles of heat treatment according to the procedures used at KHM. All samples were placed on metal trolleys placed centrally within the chamber (Fig 2). Adhesive tests were suspended from the wire grid to give a light tension. All the resin and wax samples were laid on to tissue paper on the metal shelving. FTIR analysis on samples before and after five cycles of treatment was done to assess whether Thermo-Lignum treatment could cause chemical deterioration

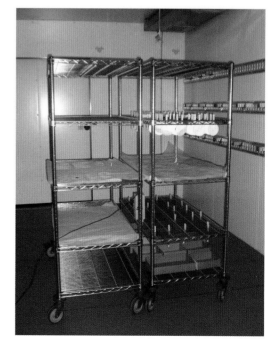

Fig 2
Test samples in the Thermo-Lignum chamber.

(oxidation, hydrolysis, and so on). FTIR was performed using an Avatar 360 ATR-FTIR spectrometer, spectra were recorded in reflection mode, from 4000 to 800cm⁻¹, 256 scans at 4cm⁻¹ resolution, using OMNIC ESP 6.1a software. Recorded spectra of blooms were compared with various commercial libraries and the Infrared and Raman Users Group (IRUG).

Results for the adhesives

Using the readings on the maximum, minimum thermometer these were found to range between 54.5°C and 60.8°C, with the first three runs recording the highest temperatures.

Visual result

On glass samples, the acrylic adhesives were prone to dimensional changes/movement indicating partial melt during heat treatment (Figs 3 and 4). Both Paraloid B72 and Paraloid B44 had spread and formed a more tendril-like dispersal, with Paraloid B44 also taking on a petrol-like hue. Paraloid B67 had spread beyond the joint, but showed no other changes. In protein-based materials there was also some indication of movement due to softening: the dispersal of the bubbles in bone glue changed, and the fish glue became tacky along the edges of the glass. In the epoxies, more bubbles appeared in the Akemi Marmokitt 1000 samples. The polyvinyl acetate became clearer, which appeared to be due to further drying of the sample. Similar changes were not observed for acrylic adhesives on wooden samples, which showed minimal changes in length or tackiness with the exception of Hxtal NYL–1 which failed during the third run. In the adhesive films on paper, there were colour changes noted to the

HMG cellulose nitrate, all of the epoxies, as well as most of the hide glues (Table 1). For many of the hide glues on paper (bone, hide and rabbit skin) the filter paper also buckled significantly with movement of the adhesive.

Length change results

For the glass samples, length increases (slippage) were observed primarily with the acrylics and hide glues (Table 1) as well as with Mowital B30H. In most cases the change in length was observed only on a single cycle, with the exception of Paraloid B72, which continued to show slippage on repeated heating. The movement of Paraloid B72 and Paraloid B67 is expected, as the temperatures reached are over 10°C higher than their Tg (approximately 40°C and 50°C respectively). Even though there was some movement with temperatures well above the Tg of many adhesives, the joints did not fail. Higher Tg acrylics showed no movement, as is the case for Paraloid B44 and Plextol (Tg 60°C and 90°C, respectively). In samples prepared on wood, no changes to the length were found, but Hxtal NYL–1 failed as the joint broke down on the first test run. The lack of slippage in wood joints is most likely due to mechanical adhesion, and indicates that failure joints with heat treatment must also consider the substrate, even if temperatures are above the Tg of the adhesive.

Weight results

All hide glues showed a decrease in weight, which is most likely a loss of water after heat cycling. Loss of water and shrinkage is also likely to be the reason that paper samples distorted. The amount of water loss in protein glues is relatively large considering that the

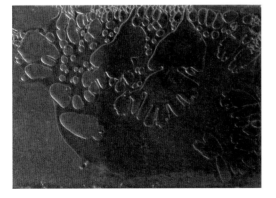
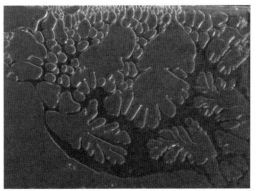

Fig 3 (far left)
Glass sample for Paraloid B72 before testing.

Fig 4 (left)
Glass sample for Paraloid B72 after testing.

percentage weight change recorded includes the weight of the glass slides. Water loss may be due to continued drying of the adhesive during heat treatment, but does indicate that damage due to water loss and dimensional change can be considerable for proteinaceous materials. Loss of water can lead to embrittlement and eventually failure of hide glue adhesives. Cellulose-based adhesives like CMC and Klucel G showed no changes due to heat treatment, so are likely not a concern when using heat treatments for pest eradication.

Samples prepared on wood and paper cannot be compared in terms of changes to the adhesive, due to the sorption of water by the substrate during heat treatment. However, in most cases the wood samples showed on average a two per cent increase in weight, with the exception of bone, hide, rabbit skin and rice starch adhered samples, which still showed a decrease in weight.

FTIR

No significant changes were noted based on ATR-FTIR analysis. However, any chemical deterioration incurred during these relatively short temperature increases are likely to be confined to the surface and not significant enough for detection on bulk samples. In some cases changes in FTIR spectra were noted, but they were not reproducible and variation within the prepared films could not be excluded.

Results for resin and waxes

Visual results

In glass samples, the shellac had become slightly granular in appearance and the benzoin more globular; the surface of the paraffin wax had become totally crizzled; whereas both shea and cocoa butters had a much more uniform appearance. These surface changes are attributed to movement due to melt. The benzoin and colophony had both darkened and the lacquer greyed in patches. On wood samples, a slight darkening had occurred to colophony and a shadowing to the lacquer. Paraffin, microcrystalline and Cosmolloid 80H waxes and shea butter all appeared to have partially melted and been absorbed into the wood. On paper samples, benzoin and colophony again showed darkening, as well as shellac, and the surface appearance of lacquer again was altered (shadowing). Most of the waxes showed some movement, and both the shea and cocoa butters had become transparent and spread to various degrees.

FTIR

Notable change was only observed for shea butter (Fig 5). Alteration appears to be due to ester formation based on the shift in the carbonyl from free fatty acid ($1709 cm^{-1}$) to $1730 cm^{-1}$ (ester) and the shift in C–O to the ester stretch

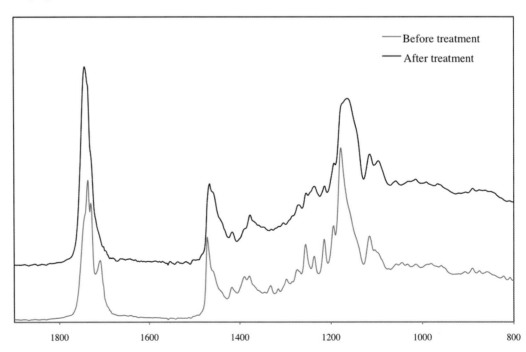

Fig 5
Fingerprint region of FTIR spectra of shea butter before and after Thermo-Lignum heat treatment.

at ~1300cm⁻¹. There were slight changes also noted in dammar and gum arabic in the OH stretching region that may indicate a gain and loss of water, respectively.

Identification of blooms

For two of the blooms, the best correlation based on FTIR is long chain free fatty acids such as stearic or palmitic acid, possibly from an emollient or oil applied to the surface (Fig 6). The band 1692cm⁻¹ is assigned to the carbonyl bonds in free fatty acids, the less intense shoulder at 1723cm⁻¹ to fatty esters (triglycerides). These blooms appear to be consistent with those studied by Pearlstein (1986), identified as fats. In the other two cases, the bloom was not identified. FTIR spectra correlated best with an unidentified conjugated aromatic, which may be a natural component of the wood.

Conclusion

In this preliminary investigation, most of the adhesives and resins showed some change due to heat treatment, including weight loss, dimensional changes, and discoloration. This raises the point that while gross visual changes to objects may not be routinely observed with the use of Thermo-Lignum heat treatment, our results indicate that there are likely changes that may be apparent upon closer visual

examination and analysis. The primary concern identified is damage to proteinaceous materials, and a moratorium on the Thermo-Lignum treatment is recommended for skins and leathers pending more detailed analyses. In common acrylics used in conservation, there was slippage, but not failure, of adhesive joints. Examination prior to treatment is needed to ensure that potential failure or distortion of these acrylics is deemed acceptable. Bloom formation on wooden objects was found to be due to migration of fats during heating. These coatings were not visible before treatment. This raises concerns for the heat treatment of objects where fats and oils may have been traditionally applied, which is the case for certain wooden sculptures and many leather artefacts. From this testing it is apparent that we need to have a full understanding of traditional manufacturing methods and the materials used in objects, together with any historic conservation treatments, before carrying out heat treatment. This preliminary investigation did not address synergistic effects, which are expected in the treatment of museum objects. Organic materials may have deterioration products that can affect other components in the artefact (corrosion of metals, discoloration of textiles) and differences in the thermal expansion coefficients between materials can result in mechanical damage.

This investigation raises the question of the optimal protocols for heat treatment when

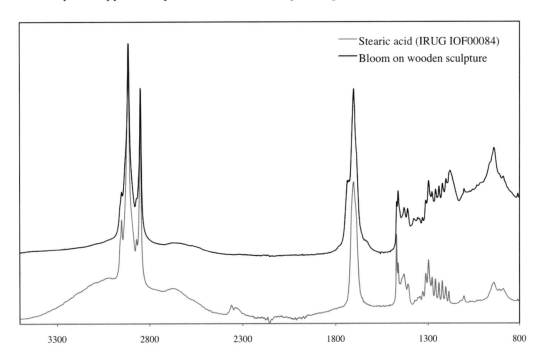

Fig 6
FTIR of bloom sampled from the surface of a wooden sculpture after heat treatment. The spectra correlate best with a long chain fatty acid such as stearic acid.

used on museum collections. The existing protocols were largely developed through the analyses of commercial lumber, and are not necessarily directly transferable to ethnographic collections where materials are often less dense or more porous than wood, and likely have different moisture sorption and thermal properties. It would be beneficial to re-examine temperature gradients and hold times needed for pest eradication in different collection materials, as current protocols may be harsher than necessary for museum applications. Collection management policies need to be addressed to determine if the time benefits outweigh the potential damage to materials occurred during treatment.

Acknowledgements

The authors would like to thank the American Scandinavian Foundation for their support, and Prof. Doug A. Loy (Department of Chemistry, University of Arizona) for useful discussions.

References

Ackery, P R, Testa, J M, Ready, P D, Doyle, A M and Pinniger, D B 2004 'Effects of high temperature pest eradication on DNA in entomological collections'. *Studies in Conservation* 49(1), 35–40

Berner, G G and Ogilvie, T M A 2005 'Thermal methods of pest eradication: Their effect on museum objects'. *The Conservator* 29, 5–18

Feller, R L 1967 'Control of the deteriorating effects of light on museum objects: Heating effects of illumination by incandescent lamps'. *Museum News* 46(9), 39–47

Feller, R L 1994 *Accelerated Ageing: Photochemical and Thermal Aspects*. Pasadena, CA: The Getty Conservation Institute

Hardboard, R 1999 'Thermal Treatment Test'. *Thermal Treatment Test Report Highlights*. Unpublished Internal Report, Natural History Museum, London

McCormick-Goodhart, M 1996 'The allowable temperature and relative humidity range for the safe use and storage of photographic materials'. *Journal of the Society of Archivists* 17(1), 7–21

Pearlstein, E 1986 'Fatty bloom on wood sculptures from Mali'. *Studies in Conservation* 31(2), 83–91

Strang, T 1992 'A review of published temperatures for the control of pest insects in museums'. *Collection Forum* 8(2), 41–67

Thermo-Lignum® WARMAIR system nd www.thermolignum.com/Warmair.html (accessed 31 August 2011)

Tscherne, F and Schachenhofer, B 2008 'Research study on Thermo Lignum® Warmair treatment on art objects with paint and gilt finishes'. Paper delivered at the *COST Action IE0601 Conference Wood Science for Conservation of Cultural Heritage, Braga, Portugal,* 5–7 November 2008

Wang, X 2010 'Heat sterilization of wood'. *Wood Handbook: Wood as an Engineering Material*. General Technical Report FPL-GTR-190. Madison, WI: U.S. Department of Agriculture, Forest Service, Forest Products Laboratory, www.fpl.fs.fed.us/documnts/fplgtr/fplgtr190/chapter_20.pdf (accessed 31 August 2011)

18

Effects of fumigants and non-chemical treatments on DNA molecules and proteins: case studies on natural history specimens and proteinaceous components of museum objects

Rika Kigawa* and Tom Strang†

*National Research Institute for Cultural Properties, Tokyo, Japan, rkigawa@tobunken.go.jp
(author for correspondence)
www.tobunken.go.jp
†Canadian Conservation Institute, Ottawa, Ontario, Canada, Tom.Strang@PCH.GC.CA
www.pch.gc.ca/cci-icc

Abstract

It has been shown that some fumigants such as methyl bromide, ethylene oxide, propylene oxide and methyl iodide can affect specimen DNA and amplifying DNA fragments from the extracted specimen DNA by PCR (polymerase chain reaction). On the other hand, sulfuryl fluoride and non-chemical methods did not have any effect on DNA. After overview of the knowledge in 10 years, we report more detail on the effects on DNA by thermal methods (heating and low temperature) with samples preconditioned to various relative humidity (RH) values. Consequently, DNA of specimens pre-equilibrated with low or moderate RH remained intact after thermal treatments. But DNA of the samples pre-equilibrated with high RH or in extreme water-soaked condition was occasionally affected. Also we briefly overview our recent work on effects of pest controlling treatments on protein-based natural specimens.

Keywords

Fumigants, non-chemical treatments, thermal treatments, natural specimens, DNA extraction, DNA analysis, PCR, proteinaceous objects

Introduction

Fumigants and other non-chemical methods have been applied to eradicate insects or moulds to stop biological damage to objects. When we choose a treatment method, as well as its efficacy to the target pest organisms, the major criterion is that the method will not damage the objects. People in western countries have been concerned for decades, 'Fumigants could damage objects, so don't use them based on some information of possible damage.' But in Japan, fumigants have been used for many years as obvious damage were not actually seen or demonstrated.

We define damage was clearly undesirable when visible alteration of objects happened. However, there is also an important category of damage where we cannot observe significant changes with the human eye. Such invisible changes are not a problem for aesthetics, but could affect chemical integrity and stability of some kinds of objects. Also, chemical alteration may change the original structure and information within the objects. The important thing is to discover precisely what happens to the objects.

As an example of invisible adverse effects, we have reported that some fumigants can affect DNA of natural history specimens (Kigawa 2003). The fumigants which we call 'alkylating agents', such as methyl bromide, methyl iodide, ethylene oxide and propylene oxide, made DNA molecules in natural specimens fragile and made subsequent DNA analysis difficult, though there were little visible adverse effects on the specimens. Since DNA analysis of natural history specimens has increasing importance in modern taxonomic studies, damage by pest treatments to DNA is a serious problem.

These significant changes to DNA prompted us to evaluate effects also on proteinaceous objects, as proteins are a major group of macromolecules that form the structure of specimens. Proteins also have chemical bonds which are commonly suspected to be altered by fumigants. Effects of fumigants and the alternative treatment measures for eradicating museum pests on proteinaceous objects are also overviewed from our recent work.

Effects on DNA analysis of natural history specimens: An overview of 10 years

In the 10 years since the conference, *2001: A Pest Odyssey*, several works have been published regarding influence of pest controlling procedures on museum specimen DNA (Kigawa *et al* 2003; Ackery *et al* 2004; Espeland *et al* 2010). Ackery *et al* (2004) reported that the Thermo Lignum® heat treatments of their entomological collections did not affect DNA analysis of the specimens. Espeland *et al* (2010) demonstrated that dichlorvos severely affected integrity of the entomological specimen DNA and subsequent amplification by PCR, especially with the high concentration of its dosage, on the other hand paradichlorobenzene and naphthalene did not. It has also been shown that many fumigants, except for sulfuryl fluoride (Vikane®), caused adverse effects on the DNA and the subsequent DNA amplification

process of model natural history specimens (Kigawa *et al* 2003).

A schematic diagram of the Kigawa *et al* testing procedure is shown in Figure 1. A specimen was divided for various treatment options to compare the results with samples of essentially the same individual origin. After treatments, two small pieces (approximately 20mg each) were cut from the treated samples to extract DNA. Taking duplicate samples for extracting DNA was to reduce the levels of experimental errors of the procedure.

We tested two kinds of specimens (muscle and mushroom) in our previous study (Kigawa *et al* 2003) and a typical example of the freeze-dried chicken muscle is shown in Figure 2. The treatment details are in Table 1. The fumigants such as methyl bromide and ethylene oxide mixed gas (ME), methyl bromide (MB), ethylene oxide (EO and EH), propylene oxide (PO), methyl iodide (MI) caused fragmentation of extracted specimen DNA, which is shown as smeared patterns in Figure 2, top. Also with the samples which showed smear patterns of extracted DNA, amplification of DNA fragments by PCR was influenced, especially when amplifying longer fragments from a single copy nuclear gene (*see* Fig 2, bottom). But the fumigant of sulfuryl fluoride (SF) and non-chemical treatment options, low temperature at –30°C (LT), heating at 60°C (HT) and carbon dioxide (CO_2), did not show any such adverse effects. Whitten *et al* (1999) also reported that SF did not affect the DNA extraction and amplification from herbarium specimens. It can be said that SF works on target insects in a different way than other fumigants tested here. Other fumigants examined (MB, EO, PO and MI) are

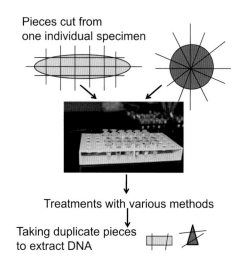

Pieces cut from
one individual specimen

Treatments with various methods

Taking duplicate pieces
to extract DNA

Fig 1
A schematic diagram of the experimental procedure.

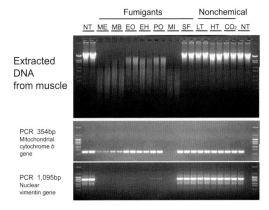

Extracted
DNA
from muscle

PCR 354bp
Mitochondrial
cytochrome *b*
gene

PCR 1,095bp
Nuclear
vimentin gene

all alkylating agents that can cause modification of DNA molecules, making covalent complexes with DNA bases to form DNA adducts. They are known as carcinogens and in this context, dichlorvos, which affected the specimen DNA (Espeland *et al* 2010), is also thought to be a putative carcinogen.

From the characteristics of the alkylating agents to modify DNA bases, we might be concerned about a possibility of occasional changes in DNA sequence data obtained from fumigated specimens. Some alkylating agents can cause alkylation of DNA at certain sequence-specific sites, which may be a problem. But if the alkylation occurs in non-specific manner,

the influence would be negligible with the bulk sequencing data obtained with large copies of PCR products. In our analysis, we sequenced some of the PCR products, but we did not observe any change in the DNA sequences with fumigated specimens, at least in our study (Kigawa *et al* 2003).

Before conducting the experiments on DNA analysis, the authors were interested in whether adverse effects can really happen by just a single treatment in the conditions usually applied in real museum communities. We did not have much concrete evidence for this. The alkylating agents are known to react *in vivo* with DNA and proteins, in the body of living organisms which have sufficient water and this is why they are used to eradicate insects or moulds. But museum objects in most case have a lower water content than vital insect or microbe pests, the target organisms by fumigants. In this sense, it was rather surprising that the fumigants affected the DNA molecules so much in the dry specimens, even by a single treatment by common procedures applied in Japanese museums. Together with such knowledge, and by the promotion of integrated pest management (IPM) policy for museums, large-scale fumigation has now become rare in Japan over the past 10 years.

Fig 2

Effects of fumigants and non-chemical methods on freeze-dried muscle specimen DNA. NT: non-treated, ME: methyl bromide/ethylene oxide mixed gas, MB: methyl bromide, EO and EH: ethylene oxide, PO: propylene oxide, MI: methyl iodide, SF: sulfuryl fluoride, LT: low temperature, HT: heating, CO_2: carbon dioxide. Details of the treatments are referred to in Table 1.

(Data edited from Kigawa et al 2003)

Table 1 Treatment conditions

Sample	Treatment	Trade name	Concentration	Treatment period	Target
NT	Non–treated	–	–	–	–
ME	Methyl bromide 86% wt/ ethylene oxide 14% wt	EKIBON	100g/m³	48 hours	Insects and moulds
MB	Methyl bromide	–	86g/m³	48 hours	Insects
EO	Ethylene oxide	–	14g/m³	48 hours	Insects
EH	Ethylene oxide 15% wt/ HFC R–134a 85% wt	EKIHUME	200g/m³	48 hours	Insects and moulds
PO	Propylene oxide	–	48g/m³	48 hours	Insects and moulds
MI	Methyl iodide	IOGUARD	120g/m³	72 hours	Insects and moulds
SF	Sulfuryl fluoride	Vikane®	50g/m³	48 hours	Insects
LT	Low temperature at –30°C	–	–	1 week	Insects
HT	Heating at 60°C	–	–	24 hours	Insects
CO_2	Carbon dioxide	–	60% volume	2 weeks	Insects

Fumigation was performed at 23 ± 1°C.

Thermal treatments and DNA: With specimens preconditioned under various RH values

Possible link to knowledge on seed viability

Strang (1999) reviewed seed viability of plant specimens in the context of thermal treatments used to control insect pests, equating lowered viability to macromolecular damage. It was concluded that, if the seeds are in well-dried conditions (specimens preconditioned below 50%RH before heating at 55–60°C, and specimens preconditioned below 14 per cent EMC dry basis before low temperature treatments down to –30°C), thermal methods would not affect seed viability (Strang 1999). Both of these conditions are normally met in controlled museum storage environments. Chalfoun and Tuross (1999) examined modern barley seeds heated in a regime from gentle to charring conditions for the preservation of both DNA and proteins, and reported that extended heating at 100°C had little effect, although higher temperature caused the chemical modification and degradation of both protein and DNA as well as the failure of immunological detection by Western blotting (to measure protein accumulation in cells) and PCR.

Though seed viability is a complex process where many molecules such as DNA and proteins are involved, at least where seed viability is well preserved, DNA molecules are also expected to be well preserved.

In our previous experiments (Kigawa *et al* 2003), we used dry freeze-dried samples and no significant problem was observed on extracted DNA by low temperature treatment or heating. However, it is possible that repeated freezing and thawing can affect DNA, if the samples are damp or wet, probably because of mechanical stress by ice formation. Also, we do not know what happens with damp or wet samples by heating. Therefore, we conducted experiments with samples pre-conditioned under various RH values.

Procedure

Freeze-dried samples of muscle from a chicken (*Gallus gallus*) and a mushroom (*Lentinus edodes*) were prepared by the same procedure as Kigawa *et al* 2003. The samples were pre-equilibrated for 24 hours under different RH conditions shown in Table 2 in glass chambers prior to thermal treatments. And the samples were put into 1.5ml polypropylene tubes and treated under low temperature at –30°C for one week or by heating at 60°C for 24 hours, then returned to room temperature and kept for 24 hours. The treated samples were dried with silica gel at room temperature for one week. DNA extraction and PCR procedure were essentially the same as Kigawa *et al* 2003.

Results

In Table 3, the results with the freeze-dried muscle specimen are summarised together with the description of visible change. The samples pre-conditioned under approximately 100%RH slightly shrank after the low temperature or heating treatments. Pre-water-soaked samples all showed heavy shrinkage, and the colour changed darker. Therefore, wet or extremely high RH showed significant visible change even before we discuss the effects on DNA.

The extracted DNA from the heat-treated samples pre-conditioned under approximately 100%RH showed a bit of smeared pattern. Also all water-soaked (wet) samples showed smeared patterns in the extracted DNA and the smeared pattern was most significant with the heat-treated sample.

Table 4 describes the results on the mushroom specimen. The mushroom samples pre-equilibrated in 100%RH or water-soaked

Table 2 Preconditioning of the samples for thermal eradication

Conditions	Details	State of samples after the RH preconditioning
Dry	Stored with dry silica gel	Dry, hard
33%RH	Stored with saturated solution of MgCl$_2$ 6H$_2$O	A bit softened
53%RH	Stored with saturated solution of Mg(NO$_3$)$_2$ 6H$_2$O	Soft
84%RH	Stored with saturated solution of KCl	Soft and elastic like dry sponge
100%RH	Stored with distilled water	Elastic, viscous
Wet	Soaked in distilled water	Wet sponge like

Pieces of approximately 150mg of freeze-dried chicken muscle and 60mg of freeze-dried mushroom were placed in glass chambers adjusted to various RH conditions for 24 hours prior to thermal treatments (-30°C for one week or 60°C for 24 hours).

Table 3 Summary of the effects on DNA extraction and amplification with PCR: Freeze-dried chicken muscle specimen

Sample	Visible change	State of extracted DNA	PCR, 354bp multi–copy gene	PCR, 1095bp single–copy gene
NT*	None	Good	+++	+++
ME*	None	Smear	+	–
MB*	None	Smear	+	–
EO*	None	Smear	++	–
EH*	None	Smear	++	–
PO*	None	Smear	++	–
MI*	None	Smear	–	–
SF*	None	Good	+++	+++
CO_2*	None	Good	+++	+++
NT, dry	None	Good	+++	+++
LT, dry	None	Good	+++	+++
HT, dry	None	Good	+++	+++
NT, 33%RH	None	Good	+++	+++
LT, 33%RH	None	Good	+++	+++
HT, 33%RH	None	Good	+++	+++
NT, 53%RH	None	Good	+++	+++
LT, 53%RH	None	Good	+++	+++
HT, 53%RH	None	Good	+++	+++
NT, 84%RH	None	Good	+++	+++
LT, 84%RH	None	Good	+++	+++
HT, 84%RH	None	Good	+++	+++
NT, 100%RH	None	Good	+++	+++
LT, 100%RH	Slight shrinkage	Good	+++	+++
HT, 100%RH	Slight shrinkage	Slightly smear	+++	++
NT, wet	Heavy shrinkage, coloured darkly	Slightly smear	++	–
LT, wet	Heavy shrinkage, coloured darkly	Slightly smear	++	–
HT, wet	Heavy shrinkage, coloured darkly	Smear	++	–

*Results summary from Kigawa *et al* (2003). NT: non-treated, ME: methyl bromide/ ethylene oxide mixed gas, MB: methyl bromide, EO and EH: ethylene oxide, PO: propylene oxide, MI: methyl iodide, SF: sulfuryl fluoride, CO2: carbon dioxide, LT: low temperature at −30°C, HT: heating at 60°C. Details of the treatments are in Tables 1 and 2. Thermal treated samples were dried under silica gel for one week after the treatments. +++: sound amplification, ++: enough amplification, +: visible band amplified, -: no or little amplification.

shrank dramatically, accompanied with some weight loss.

In this case, extracted DNA of the heat-treated sample pre-conditioned in 84%RH showed a bit of smeared pattern (Table 4 and Fig 3, top). Also the extracted DNA from the thermal treated samples pre-conditioned in 100%RH showed smeared patterns (*see* Fig 3,

Table 4 Summary of the effects on DNA extraction and amplification with PCR: freeze-dried mushroom specimen

Sample	Visible change	State of extracted DNA	PCR, 310bp multi-copy gene	PCR, 987bp single-copy gene
NT*	None	Good	+++	+++
ME*	None	Smear	+++	−
MB*	None	Smear	+++	−
EO*	None	Smear but the genomic DNA band remained	+++	+
EH*	None	Smear	+++	−
PO*	None	Smear but the genomic DNA band remained	+++	+
MI*	None	Smear	+++	−
SF*	None	Good	+++	+++
CO_2*	None	Good	+++	+++
NT, dry	None	Good	+++	+++
LT, dry	None	Good	+++	+++
HT, dry	None	Good	+++	+++
NT, 33%RH	None	Good	+++	+++
LT, 33%RH	None	Good	+++	+++
HT, 33%RH	None	Good	+++	+++
NT, 53%RH	None	Good	+++	+++
LT, 53%RH	None	Good	+++	+++
HT, 53%RH	None	Good	+++	+++
NT, 84%RH	None	Good	+++	+++
LT, 84%RH	None	Good	+++	+++
HT, 84%RH	Slight shrinkage	Slightly smear	+++	+++
NT, 100%RH	Heavy shrinkage	Smear	+++	++
LT, 100%RH	Heavy shrinkage	Smear	+++	++
HT, 100%RH	Heavy shrinkage	Smear	+++	++
NT, wet	Heavy shrinkage	Totally degraded	+++	−
LT, wet	Heavy shrinkage	Totally degraded	++	−
HT, wet	Heavy shrinkage	Totally degraded	++	−

*Results summary from Kigawa et al (2003). NT: non-treated, ME: methyl bromide/ ethylene oxide mixed gas, MB: methyl bromide, EO and EH: ethylene oxide, PO: propylene oxide, MI: methyl iodide, SF: sulfuryl fluoride, CO_2: carbon dioxide, LT: low temperature at −30°C, HT: heating at 60°C. Details of the treatments are in Tables 1 and 2. Thermal treated samples were dried under silica gel for one week after the treatments. +++: sound amplification, ++: enough amplification, +: visible band amplified, -: no or little amplification.

top). Water soaking itself affected significantly the integrity of specimen DNA and PCR amplification of a 987bp fragment.

From the observation of the conditions of extracted DNA, it is recommended to keep specimens of these kinds in dry and moderate RH conditions before treating them by thermal

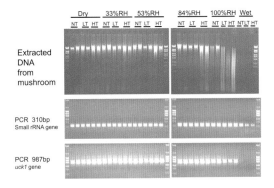

Extracted
DNA
from
mushroom

PCR 310bp
Small rRNA gene

PCR 987bp
uck1 gene

methods, especially by heating at around 60°C. But, as a whole, DNA amplification by PCR tested here was not interfered with unless the specimens were wet.

Effects on proteinaceous objects

The example of fumigant-induced change to DNA prompted us to evaluate effects on proteinaceous objects, as proteins are also major macromolecules of biological origin that have chemical bonds that are possibly attacked by some fumigants.

There are some references to the effects of fumigants on proteinaceous materials (Dawson 1988; Baker *et al* 1990; Florian 1988). However, there is not much published experimental evidence to show clear chemical changes in proteinaceous objects by practical treatments for pest eradication in the field of conservation. Our interest is to determine whether we could observe deleterious effects on proteinaceous objects by typical conditions as commonly applied to museum objects.

A brief summary of our recent work (Kigawa *et al* submitted) is shown in Table 5. We adopted methods of protein electrophoresis, Fourier transform infrared attenuated total reflection (FTIR-ATR), differential scanning calorimetry (DSC), thermal gravimetric analysis (TGA), thermal microscopy (Tmic) and amino acid analysis.

As a result, though there was no visible change on the specimen, some fumigants, methyl bromide, methyl iodide and sulfuryl fluoride caused modification and denaturation of the freeze-dried muscle proteins. But as the change was not the kind that makes the specimen fragile immediately, it is difficult to define the damage in the case of the muscle specimen. At least, when we wish to conserve the

original forms of muscle proteins in the specimens, it can be said to be an adverse effect. Such fumigants would also affect the results of future analytical studies. This means that records of historical treatments by fumigants are important when we consider future research.

Regarding animal glue, we did not observe significant changes in our study. Since animal glues are subjected to heat, water and chemical reagents (lime, sulphates, and so on) in their production process, they may be in a more stable form when they are supplied as products.

We did not detect any significant changes in silk textiles. Silk proteins have a structural function to protect insect cocoons and, from this natural selection, silk proteins might be resistant to chemical agents.

Our study looked at results of a single treatment; therefore effects of repeated fumigant treatments should be considered in future studies.

Conclusion

Significant degradation of extracted DNA of the specimens has been shown in the samples treated by the fumigants; methyl bromide, methyl bromide/ethylene oxide mixed gas, ethylene oxide, propylene oxide and methyl iodide, even with a single fumigation. Subsequent amplification of DNA fragments by PCR was difficult in these samples, especially in amplifying larger DNA fragments. Also from detailed study of effects on DNA by thermal treatments, it is recommended that specimens

Fig 3
Effects of thermal treatments on DNA of mushroom specimen equilibrated to various RH conditions. NT: non-treated, LT: low temperature at minus 30°C, HT: heating at 60°C. Details of preconditioning are shown in Table 2. For some samples that shrank significantly after the treatments, we were unable to take duplicate 20mg pieces for DNA extraction.

Table 5 Brief summary of effects on proteinaceous materials

Sample	Effects
Freeze-dried chicken muscle	No visible change from any of the treatments. Evidence of some covalent bond modification and denaturation of proteins by some fumigants such as methyl bromide, methyl iodide and sulfuryl fluoride.
Animal glue	No visible change from any of the treatments. No significant evidences of change so far.
Silk textiles	No visible change from any of the treatments. No significant evidences of change so far.

Results of testing treatments: fumigation with methyl bromide, methyl iodide, ethylene oxide, propylene oxide and sulfuryl fluoride, low temperature treatment at −30°C for one week, heat treatment at 47 to 60°C for 24 hours, low oxygen treatment for three weeks, carbon dioxide treatment for two weeks (from Kigawa *et al* submitted).

are kept under dry to moderate RH conditions before thermal treatments.

Proteins appear to respond differently to fumigants, based on their structure and characteristics. Freeze-dried muscle and soft tissue proteins appear to be more susceptible to fumigant alteration than highly processed animal glue collagen and silk textiles.

Acknowledgements

The authors would like to express special thanks to Yoshiko Miyazawa and TAKARA BIO INC. for their technical assistance. The authors also gratefully acknowledge: Noriko Hayakawa and Gregory Young for their great cooperation in protein studies, and Akiko Okimoto, Akemi Toyoda and Shigeo Aoki for their kind assistance in sample preparation. This work was supported in part by Grants-in-Aid for Scientific Research (Kakenhi) from Japan Society for the Promotion of Science (JSPS). Finally, we express our special gratitude to David Pinniger for his warm encouragement to write a paper for this conference and his critical reading of the manuscript.

References

Ackery, P R, Testa, J M, Ready, P D, Doyle, A M and Pinniger, D B 2004 'Effects of high temperature pest eradication on DNA in entomological collections'. *Studies in Conservation* **49**, 35–40

Baker, M T, Burgess, H D, Binnie, N E, Derrick, M and Druzik, J R 1990 'Laboratory investigation of the fumigant Vikane® in Grimstad, K (ed) *9th Triennial Meeting, Dresden, German Democratic Republic, 26–31 August 1990 Preprints*, Los Angeles: ICOM Committee for Conservation, 804–11

Chalfoun, D J and Tuross, N 1999 'Botanical remains: Utility in protein and DNA research'. *Ancient Biomolecules* **3**, 67–79

Dawson, J 1988 'The effects of insecticides on museum artifacts and materials' *in* Zycherman, L A and Schrock, J R (eds) *A Guide to Museum Pest Control*. Washington, DC: Foundation of the American Institute for Conservation of Historic and Artistic Works and Association of Systematics Collections, 135–50

Espeland, M, Irestedt, M, Johanson, K A, Akerlund, M, Bergh, J E and Källersjö, M 2010 'Dichlorvos exposure impedes extraction and amplification of DNA from insects in museum collections'. *Frontiers in Zoology* **7**(2). www.frontiersinzoology.com/content/7/1/2 (accessed 19 September 2011)

Florian, M-L 1988 'Ethylene oxide fumigation: A literature review of the problems and interactions with materials and substances in artifacts', *in* Zycherman, L A and Schrock, J R (eds) *A Guide to Museum Pest Control*, Washington, DC: Foundation of the American Institute for Conservation of Historic and Artistic Works, Association of Systematics Collections, 151–8

Kigawa, R, Nochide, H, Kimura, H and Miura, S 2003 'Effects of various fumigants, thermal methods and carbon dioxide treatment on DNA extraction and amplification: A case study on freeze-dried mushroom and freeze-dried muscle specimens'. *Collection Forum* **18**(1–2), 74–89

Kigawa, R, Strang, T, Hayakawa, N, Kimura, H and Young, G (Submitted) 'Investigation of effects of fumigants on proteinaceous components of museum objects (muscle, animal glue and silk) in comparison with other non-chemical pest eradicating measures'.

Strang, T J K 1999 'Sensitivity of seeds in herbarium collections to storage conditions and implications for thermal insect pest control methods' *in* Metsger, D A and Byers, S C (eds) *Managing the Modern Herbarium: An Interdisciplinary Approach*. Vancouver: Elton–Wolf, 81–102

Whitten, W M, Williams, N H and Glover, K V 1999 'Sulphuryl fluoride fumigation: Effect on DNA extraction and amplification from herbarium specimens'. *TAXON (Journal of the International Association for Plant Taxonomy)* **48**(3), 507–10.

19

Integrated pest management at the National Museum of Ethnology, Japan: re-evaluation of preventive measures and control strategies

Naoko Sonoda* and Shingo Hidaka†

*National Museum of Ethnology, Osaka, Japan, sonoda@idc.minpaku.ac.jp
(author for correspondence)
†National Museum of Ethnology, Osaka, Japan, s-hidaka@idc.minpaku.ac.jp
www.minpaku.ac.jp/english/

Abstract

The National Museum of Ethnology, Japan houses an extensive and growing collection of artefacts from around the world. To safeguard this collection, a series of preventive and control measures have been undertaken as part of an integrated pest management programme (IPM). With regards to preventive measures, two specially customised computer programs have been developed and are currently in use, namely: a pest monitoring data analysis system, and a temperature/relative humidity (RH) data analysis system. This paper describes a systematic review of the data conducted using these systems. The former system is applied to the analysis of seasonal and geographic pest trends and to the evaluation of the effectiveness of IPM measures. The latter system aids an understanding of the museum environment and an evaluation of climate control levels. The pest control policy is then outlined and information is given on technical developments in the field, including *in situ* heat treatment of large objects and freezing in a specially made heating/freezing chamber.

Keywords

Integrated pest management, IPM, preventive conservation, traps, climate control, heat treatment, freezing

Introduction

The National Museum of Ethnology, Japan was established in 1974 as an inter-university research institute. The museum has an extensive and growing collection of artefacts from around the world. As many of the artefacts are composed of organic materials susceptible to bio-deterioration, the museum implements a series of preventive and control measures as part of an integrated pest management (IPM) programme (Sonoda and Hidaka 2008).

Two customised computer programs have been developed to facilitate the daily running of the museum. One is a pest monitoring data analysis system that rapidly identifies problem areas or species and permits the visualisation of geographic distribution and concentration of pests. The other, a temperature/relative humidity (RH) data analysis system, aids an understanding of the museum environment. We describe here a systematic review of data that was conducted using these systems in an evaluation of the effectiveness of IPM and the appropriateness of current climate control levels. In the final part of this paper, we present our pest control policy alongside information on technical developments, including *in situ* heat treatment of large objects and freezing in a specially made heating/freezing chamber.

Pest monitoring data analysis

Seasonal and geographic pest trends monitoring using insect traps started in the museum in 1992. Two kinds of insect trap are set at fixed points in the storage zone, the gallery zone and in the other zone (comprising the rooms adjacent to the storage zone) for two weeks, and then collected. As of February 2011, 304 sheet traps and 181 pheromone traps for cigarette beetles are in use.

The pest monitoring data analysis system was developed in 2004 and revised in 2007 using the online software PowerOLAP (PARIS Technologies Inc USA). After each investigation, the system is currently used to identify problem areas or species and permits visualisation of geographic distribution and concentration of pests through an original mapping based on Microsoft Excel. Particular attention is paid to the occurrence of booklice and/or of silverfish as their presence can be interpreted as an indicator of high RH, and consequently a higher risk of mould growth.

Seasonal and geographic pest trends are examined by reviewing the data over 18 years of investigations. Some of our findings are described below.

Pests are most frequently captured in summer, followed in order by spring, autumn, and winter. There are no clear seasonal characteristics in the species captured, probably because the museum environment is controlled for human comfort, and consequently for pest comfort, throughout the year. Table 1 shows that, in the storage zone, constant attention should be paid to booklice, beetles and moths. In the gallery zone and other zone, pests originating from outside the museum are frequently captured: flies in all seasons; spiders in spring and winter, and woodlice in summer and autumn. Booklice always rank among the most frequently captured species, except in the gallery and other zones in spring, which is a dry season in Osaka.

In the period 1992–9, the most frequently captured species were booklice and beetles, the latter becoming more prevalent around 1997–9. Between 1999 and 2000 there was a local outbreak of moths, and from 2001 a marked increase in the number of flies. At the beginning of the investigation, booklice were found mainly in the storage and gallery zones, however they are now observed mainly in the other zone. Beetles are captured mainly in the storage and gallery zones, although their numbers have clearly decreased in recent years. Moths were found almost exclusively in the storage zone in 1999–2000, during work to replace air conditioning ducts. Flies have appeared frequently in the other zone since 2001. The occurrence of pests has changed location from the storage and gallery zones to the other zone. This change occurred in 2001 when we started to strengthen communication between different sections following the sudden moth attack.

Evaluation of the effectiveness of IPM preventive measures

Our analysis revealed that, in the storage zone, pest attacks tend to spread throughout the space, probably because of the high density of artefacts there (Fig 1). Another possible explanation is that, as the storage rooms are less visited, early detection is somewhat difficult. After the moth attack of 1999–2000, we started to review past data and realised the importance of communication between different sections. It is noteworthy that the number of pests captured has decreased dramatically since better communication was established. From 2007, in order to maintain good storage conditions, we have made it a rule to clean the storage rooms and give each artefact a cursory inspection at least once a year.

In the gallery zone, pest attacks tend to concentrate on specific artefacts, such as the

Table 1 The three most captured species by zone and season

	Spring	Summer	Autumn	Winter
Storage zone	Moth and butterfly	Booklouse	Booklouse	Beetle, weevil
	Fly	Beetle, weevil	Beetle, weevil	Booklouse
	Booklouse	Moth and butterfly	Moth and butterfly	Moth and butterfly
Gallery zone	Fly	Beetle, weevil	Fly	Fly
	Beetle, weevil	Booklouse	Booklouse	Booklouse
	Spider	Fly	Beetle, weevil	Spider
Other zone	Fly	Fly	Fly	Fly
	Spider	Booklouse	Booklouse	Booklouse
	Beetle, weevil	Woodlouse, pill bug	Woodlouse, pill bug	Spider

Storage zone

- ■ Beetle, Weevil
- ◫ Bristletail
- ⊠ Termite, White ant
- ◖ Thrip
- ⊡ Earwig
- ▨ Spider
- ▨ Centipede

- ◻ Moth and Butterfly
- ▨ Booklouse
- ▨ Fly
- ⊠ Truebug
- ▨ Other Insects
- ■ Woodlouse, Pill bug
- ◻ Others

- ▨ Cockroach
- ◻ Orthopteron
- ▨ Sawfly, Wasp, Bee, Ant
- ▨ Springtail
- ◻ Mite, Tick
- ◪ Millipede

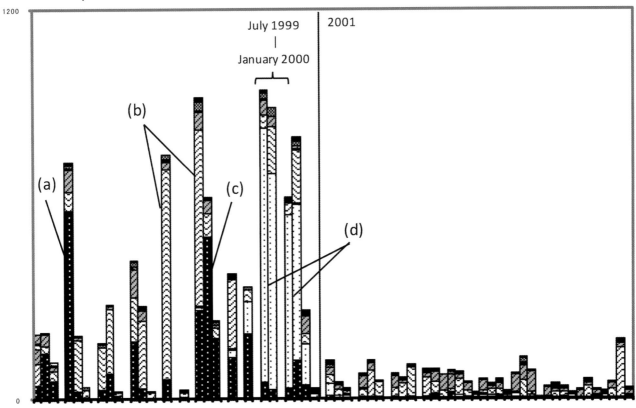

Malaysian houseboat or Mongolian tent. Therefore, in 2004 we introduced a daily inspection and a weekly thorough inspection of the artefacts prone to insect attack. This IPM inspection is performed during the routine rounds before the museum opens to the public. Figure 2 indicates that the situation improved on two occasions, in 2001 and 2004, confirming the importance of better communication and daily attention.

In the other zone, the total number of pests increased, even after the improvements made in 2001, due to an increase in the number of flies present. However, it is important to note that six traps and then a further 10 traps were added, respectively, in 2001 and in 2004 following building extension work, and since 2005 an investigation has been carried out every season, adding a spring investigation not systematically performed before. Considering these facts, the data on total pest numbers were re-examined: the number of pests caught had increased, but the actual number of pests present in each trap had decreased. Figure 3 shows the number of flies caught by the traps in the same season (spring), before and after

Fig 1
Pests captured in the storage zone from 1992 to 2010. (Note: Storage rooms 1, 2 and 3 are connected.)
(a) Beetle (general storage room 5). (b) Booklouse (general storage rooms 2, 3). (c) Beetle (general storage room 5). (d) Moth (general storage rooms 1, 2, 3).

Gallery zone

- ■ Beetle, Weevil
- ⊞ Bristletail
- ⊠ Termite, White ant
- ◪ Thrip
- ▣ Earwig
- ▨ Spider
- ▨ Centipede

- ▢ Moth and Butterfly
- ▨ Booklouse
- ▨ Fly
- ⊠ Truebug
- ▨ Other Insects
- ■ Woodlouse, Pill bug
- ▢ Others

- ▢ Cockroach
- ▢ Orthopteron
- ▨ Sawfly, Wasp, Bee, Ant
- ▨ Springtail
- ▢ Mite, Tick
- ▢ Millipede

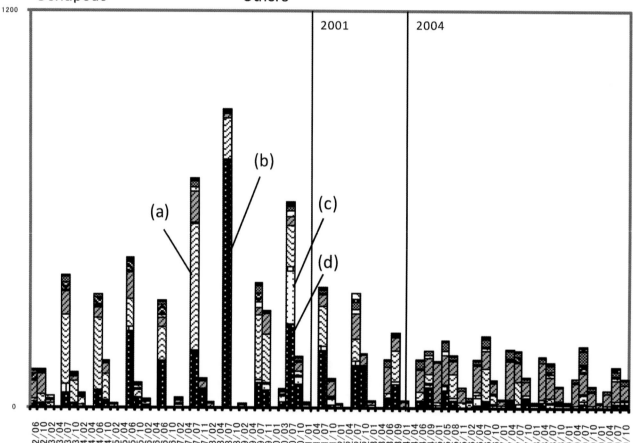

Fig 2
Pests captured in the gallery zone from 1992 to 2010. (a) Booklouse (Malaysian houseboat). (b) Beetle (saddles). (c) Moth (Mongolian tent). (d) Beetle (Mongolian tent).

replacement of a screen door on the shutter gate in February 2007.

Temperature/RH data analysis

Different levels of climate control and temperature/RH settings

Since its inception, the museum has adopted three levels of climate control: for the special storage rooms, the general storage rooms, and the exhibition galleries (Table 2). As the museum is an inter-university research institute, researchers are allowed to enter the storage rooms for research and perusal, and the storage environment is maintained for proper conservation of the collection as well as for human comfort.

Special storage rooms: Artefacts that require extra care are stored in the special storage rooms. Each room is constantly set to a specific

126

temperature and RH all year round, as appropriate to the nature of the materials stored.

General storage rooms: The collection is composed of diverse materials, and therefore the general storage rooms in which 90 per cent of the artefacts stored are set to average climate conditions: temperature is set at 26°C (±2°C) in summer and at 20°C (±2°C) in winter, varying 0.5°C per week in the transition periods, and RH is set at 52 per cent (±5 per cent) RH all year round. These rooms are air-conditioned daily between 08:20 and 18:00.

The gallery zone: The temperature is set at 26°C (±2°C) in summer and at 22°C (±2°C) in winter, varying 0.5°C per week during the transition periods. RH setting remains at 50 per cent (±10 per cent) RH all year round. The zone is air-conditioned from 09:00 to 18:00 except on Wednesdays when the museum is closed.

Since 2006, a climate control meeting has been held every two weeks to share information between conservation scientists, staff from the three sections responsible of the storage rooms (for artefacts, audio-visual materials, and rare books, respectively), staff from the building maintenance section, and members of the company working for the central operation room.

Evaluation of climate control level

Temperature/RH monitoring with data loggers was introduced gradually from 2004 to complement the conventional thermo-hydrograph recording and monitoring by the central operation room. In order to evaluate the actual level of climate control, rapid analysis was conducted using a temperature/RH data analysis system developed in 2009. Taking the annual data for 2010 (recorded at 30-minute intervals), average (m) and standard deviation (δ) of temperature and RH were first calculated for the periods in which climate settings remain constant, namely, throughout the year in the special storage rooms, and in summer and winter in the general storage rooms and exhibition galleries (see Table 2 and Figs 4 and 5). Then, a similar analysis was conducted for the transition periods, with calculation being performed for each week (Tables 3 and 4). The analysis showed that:

The actual temperature averages were found to be approximately within the settings, with the exception, in winter, of galleries 1, 5,

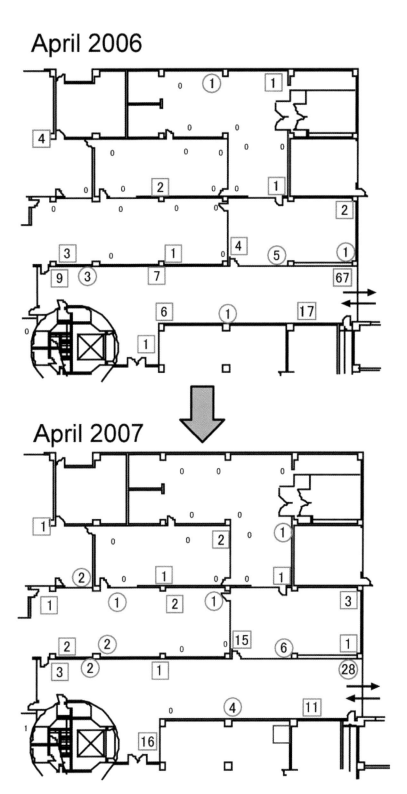

Fig 3
Mapping of flies before and after replacement of a screen door on the shutter gate.

Table 2 Temperature/RH setting, average, standard deviation (SD) and tentative control level classification for the special storage rooms (all seasons), general storage rooms (winter, summer) and exhibition galleries (winter, summer) (data 2010)

			Temperature (°C)				Relative humidity (%RH)			
			Setting	Average	SD	Classification	Setting	Average	SD	Classification
Special storage rooms	Weapons	all seasons	22 ± 1	22.2	1.33	B	50 ± 2	49.3	2.97	B
	Fur and skins	all seasons	22 ± 1	21.0	0.54	A	50 ± 2	53.2	1.08	A
	Lacquerware	all seasons	22 ± 1	23.2	1.67	B	60 ± 2	63.0	0.83	A
	Carpets	all seasons	20 ± 1	20.3	0.68	A	50 ± 5	54.0	0.89	A
	Textiles	all seasons	22 ± 1	21.8	0.24	A	55 ± 2	53.8	0.44	A
	Original films	all seasons	12 ± 2	10.7	0.10	A	40 ± 2	44.5	0.89	A
	Films	all seasons	18 ± 2	17.4	0.23	A	45 ± 2	48.1	0.98	A
General storage rooms	2	summer	26 ± 2	25.5	0.45	A	52 ± 5	56.1	1.10	A
		winter	20 ± 2	19.9	0.41	A		50.0	0.82	A
	3	summer	26 ± 2	26.4	0.60	A	52 ± 5	53.2	0.79	A
		winter	20 ± 2	19.2	0.66	A		51.9	1.05	A
	5	summer	26 ± 2	26.8	1.42	B	52 ± 5	54.8	3.38	B
		winter	20 ± 2	18.5	1.74	B		58.9	1.35	A
	6	summer	26 ± 2	25.0	0.76	A	52 ± 5	55.1	2.19	A
		winter	20 ± 2	18.9	0.54	A		52.6	0.86	A
	7	summer	26 ± 2	25.7	0.66	A	52 ± 5	55.9	1.62	A
		winter	20 ± 2	18.1	0.51	A		53.0	2.10	A
Exhibition galleries	1	summer	26 ± 2	27.3	1.65	B	50 ± 10	53.2	4.18	B
		winter	22 ± 2	19.3	2.07	B		48.4	7.00	C
	3	summer	26 ± 2	27.6	1.55	B	50 ± 10	54.4	4.67	B
		winter	22 ± 2	21.1	2.28	B		54.1	5.32	C
	5	summer	26 ± 2	27.2	2.08	B	50 ± 10	49.3	3.51	B
		winter	22 ± 2	17.3	2.31	B		48.7	4.23	B
	7	summer	26 ± 2	26.8	1.60	B	50 ± 10	53.4	6.56	C
		winter	22 ± 2	19.1	1.62	B		51.1	6.15	C
	8	summer	26 ± 2	27.0	2.04	B	50 ± 10	53.3	4.71	B
		winter	22 ± 2	18.0	2.35	B		49.5	5.29	C

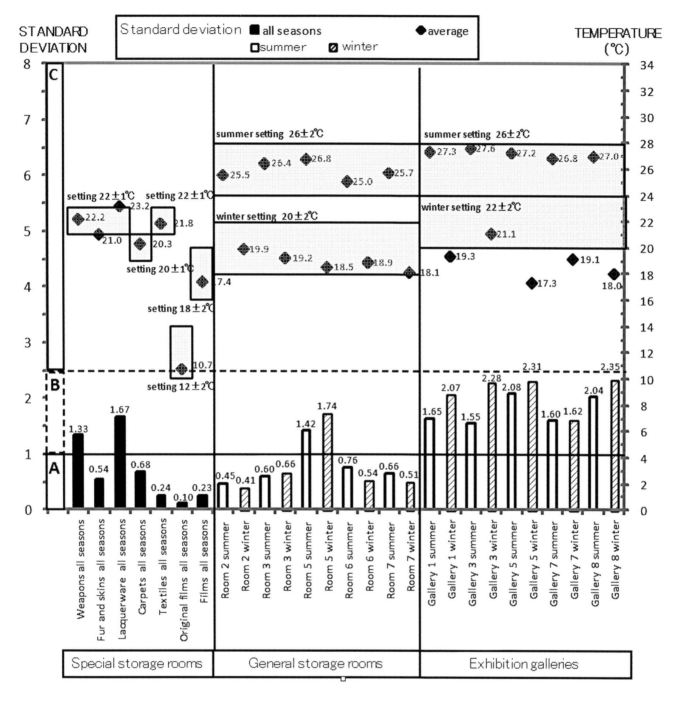

Fig 4
Temperature average (right axis) and standard deviation (left axis) for the storage rooms and galleries (data 2010).

7 and 8; their values were below the minimal set-point of 20°C and revealed to be more energy efficient. The actual RH averages were not within RH settings for the special storage rooms for fur and skins, lacquerware, original films, and films, which suggests that the allowable margin of ±2%RH is too severe. The deviation from the allowed margin is, however, negligible except for the special storage room for original films, requiring us to review the case and take necessary measures.

We attempted to classify the actual climate data into three categories, assuming tentatively that 95 per cent of the data are expected to be within the limits of $m \pm 2\delta$.

1. Class A – Temperature fluctuation within ±2°C ($0 < \delta \leq 1$)

   RH fluctuation within ±5%RH ($0 < \delta \leq 2.5$)

2. Class B – Temperature fluctuation within ±5°C ($1 < \delta \leq 2.5$)

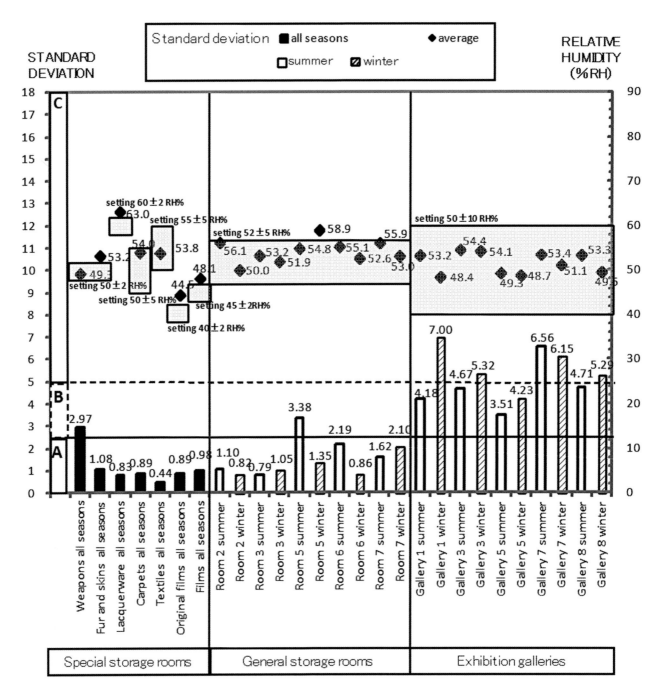

Fig 5
RH average (right axis) and standard deviation (left axis) for the storage rooms and galleries (data 2010).

RH fluctuation within ±10%RH (2.5 < δ ≤ 5)

3. Class C – Temperature fluctuation >5°C (2.5 < δ)

RH fluctuation >10%RH (5 < δ)

In summary, temperature control is performed better than RH control, and is better controlled in summer than in winter. The storage rooms are well controlled (temperature fluctuation within ±2°C, RH fluctuation within ±5%RH) in all seasons, with the exception of the special storage room for weapons and the general storage room 5. The former is a transition room between the special storage rooms and general storage rooms, and is easily influenced by the climate conditions of the general storage rooms; the latter is the only storage room situated upstairs and is not surrounded by other rooms. The special storage room for lacquerware is deliberately not temperature controlled to avoid RH fluctuation, which

Table 3 Temperature/RH setting, average, standard deviation and tentative control level classification for the general storage rooms (transition periods) (data 2010)

			Temperature (°C)				Relative humidity (%RH)			
			Setting	Average	SD	Classification	Setting	Average	SD	Classification
General storage room 2	Transition period (spring)	week 1	20.5 ± 2	20.4	0.26	A	52 ± 5	50.3	0.55	A
		week 2	21 ± 2	20.8	0.22	A		51.2	0.91	A
		week 3	21.5 ± 2	20.8	0.24	A		51.1	0.54	A
		week 4	22 ± 2	21.4	0.41	A		51.0	0.66	A
		week 5	22.5 ± 2	22.5	0.19	A		53.1	0.60	A
		week 6	23 ± 2	22.6	0.15	A		52.1	0.67	A
		week 7	23.5 ± 2	23.0	0.23	A		53.5	1.16	A
		week 8	24 ± 2	23.1	0.19	A		54.0	1.23	A
		week 9	24.5 ± 2	23.8	0.40	A		52.0	0.38	A
		week 10	25 ± 2	24.7	0.24	A		52.8	0.38	A
		week 11	25.5 ± 2	24.9	0.11	A		54.5	0.53	A
		week 12	26 ± 2	25.0	0.12	A		55.1	0.14	A
	Transition period (autumn)	week 1	25.5 ± 2	24.9	0.11	A	52 ± 5	54.0	0.25	A
		week 2	25 ± 2	24.9	0.08	A		54.2	0.64	A
		week 3	24.5 ± 2	24.6	0.12	A		52.6	0.21	A
		week 4	24 ± 2	23.8	0.42	A		51.7	0.92	A
		week 5	23.5 ± 2	22.8	0.31	A		49.0	1.50	A
		week 6	23 ± 2	22.1	0.25	A		48.2	0.81	A
		week 7	22.5 ± 2	21.6	0.25	A		48.9	1.00	A
		week 8	22 ± 2	21.5	0.17	A		49.9	0.42	A
		week 9	21.5 ± 2	21.2	0.17	A		49.9	0.52	A
		week 10	21 ± 2	20.8	0.24	A		50.0	0.46	A
		week 11	20.5 ± 2	20.3	0.25	A		49.5	0.51	A
		week 12	20 ± 2	20.0	0.17	A		49.0	0.29	A
General storage room 3	Transition period (spring)	week 1	20.5 ± 2	20.1	0.40	A	52 ± 5	51.1	0.55	A
		week 2	21 ± 2	20.7	0.35	A		51.7	0.75	A
		week 3	21.5 ± 2	20.9	0.46	A		51.1	0.85	A
		week 4	22 ± 2	21.6	0.60	A		50.5	0.79	A
		week 5	22.5 ± 2	23.0	0.26	A		51.2	0.50	A
		week 6	23 ± 2	22.8	0.33	A		51.5	0.89	A
		week 7	23.5 ± 2	23.4	0.35	A		51.6	0.73	A
		week 8	24 ± 2	23.5	0.46	A		52.6	0.97	A
		week 9	24.5 ± 2	24.6	0.62	A		51.3	0.74	A
		week 10	25 ± 2	25.5	0.40	A		51.6	0.57	A
		week 11	25.5 ± 2	25.6	0.29	A		52.6	0.66	A
		week 12	26 ± 2	25.7	0.25	A		53.3	0.50	A

Table 3 Continued

	Transition period (autumn)	week 1	25.5 ± 2	25.2	0.20	A	52 ± 5	53.5	0.39	A	
		week 2	25 ± 2	25.3	0.22	A		53.6	0.40	A	
		week 3	24.5 ± 2	24.8	0.21	A		53.4	0.34	A	
		week 4	24 ± 2	23.7	0.45	A		53.2	0.71	A	
		week 5	23.5 ± 2	22.7	0.45	A		51.8	1.22	A	
		week 6	23 ± 2	22.0	0.39	A		50.4	0.72	A	
		week 7	22.5 ± 2	21.4	0.43	A		50.6	0.63	A	
		week 8	22 ± 2	21.1	0.32	A		51.4	0.61	A	
		week 9	21.5 ± 2	20.5	0.32	A		51.1	0.58	A	
		week 10	21 ± 2	20.1	0.38	A		51.1	0.64	A	
		week 11	20.5 ± 2	19.6	0.47	A		51.1	0.71	A	
		week 12	20 ± 2	19.4	0.51	A		50.9	0.96	A	
General storage room 5	Transition period (spring)	week 1	20.5 ± 2	21.1	1.34	B	52 ± 5	57.7	1.20	A	
		week 2	21 ± 2	21.5	1.86	B		58.2	1.35	A	
		week 3	21.5 ± 2	20.5	0.81	A		58.3	0.70	A	
		week 4	22 ± 2	21.3	0.79	A		57.3	0.94	A	
		week 5	22.5 ± 2	22.1	0.48	A		58.7	1.55	A	
		week 6	23 ± 2	21.4	0.53	A		57.5	1.08	A	
		week 7	23.5 ± 2	22.4	0.47	A		58.3	1.85	A	
		week 8	24 ± 2	22.1	0.47	A		58.8	2.16	A	
		week 9	24.5 ± 2	23.8	0.85	A		56.0	1.06	A	
		week 10	25 ± 2	24.9	0.73	A		55.7	1.30	A	
		week 11	25.5 ± 2	25.1	0.35	A		57.4	1.83	A	
		week 12	26 ± 2	25.0	0.20	A		57.7	1.91	A	
	Transition period (autumn)	week 1	25.5 ± 2	25.8	0.71	A	52 ± 5	54.9	1.24	A	
		week 2	25 ± 2	25.7	0.55	A		54.7	1.10	A	
		week 3	24.5 ± 2	24.8	0.66	A		52.8	0.77	A	
		week 4	24 ± 2	22.9	1.19	B		52.5	1.29	A	
		week 5	23.5 ± 2	22.6	1.22	B		50.4	1.85	A	
		week 6	23 ± 2	22.1	1.28	B		53.0	2.39	A	
		week 7	22.5 ± 2	20.1	1.32	B		57.8	1.02	A	
		week 8	22 ± 2	19.6	0.86	A		58.5	0.66	A	
		week 9	21.5 ± 2	19.4	1.07	B		58.2	0.80	A	
		week 10	21 ± 2	19.2	1.17	B		58.4	0.83	A	
		week 11	20.5 ± 2	18.7	1.45	B		58.5	1.09	A	
		week 12	20 ± 2	19.0	1.54	B		57.8	0.86	A	
General storage room 6	Transition period (spring)	week 1	20.5 ± 2	19.0	0.53	A	52 ± 5	53.7	0.94	A	
		week 2	21 ± 2	19.5	0.52	A		54.1	1.47	A	
		week 3	21.5 ± 2	19.6	0.55	A		54.7	1.15	A	
		week 4	22 ± 2	19.8	0.59	A		54.9	1.22	A	
		week 5	22.5 ± 2	20.6	0.63	A		56.5	1.24	A	

Table 3 Continued

		week 6	23 ± 2	21.0	0.60	A		55.3	1.77	A
		week 7	23.5 ± 2	21.5	0.69	A		55.4	1.48	A
		week 8	24 ± 2	21.9	0.58	A		55.6	1.72	A
		week 9	24.5 ± 2	22.2	0.67	A		55.1	1.32	A
		week 10	25 ± 2	23.2	0.89	A		55.0	1.42	A
		week 11	25.5 ± 2	24.0	0.75	A		55.7	1.34	A
		week 12	26 ± 2	23.9	0.69	A		56.1	1.42	A
	Transition period (autumn)	week 1	25.5 ± 2	24.3	0.30	A	52 ± 5	54.7	1.20	A
		week 2	25 ± 2	23.8	0.18	A		55.0	1.21	A
		week 3	24.5 ± 2	23.5	0.22	A		54.2	0.81	A
		week 4	24 ± 2	23.1	0.29	A		53.3	1.56	A
		week 5	23.5 ± 2	22.7	0.37	A		50.6	1.94	A
		week 6	23 ± 2	22.1	0.34	A		49.3	0.83	A
		week 7	22.5 ± 2	21.8	0.43	A		49.1	0.72	A
		week 8	22 ± 2	21.4	0.28	A		49.4	0.80	A
		week 9	21.5 ± 2	21.0	0.27	A		49.2	0.65	A
		week 10	21 ± 2	20.6	0.25	A		49.5	0.91	A
		week 11	20.5 ± 2	20.5	0.47	A		52.0	0.99	A
		week 12	20 ± 2	20.6	0.57	A		52.4	0.69	A
General storage room 7	Transition period (spring)	week 1	20.5 ± 2	18.6	0.33	A	52 ± 5	52.8	0.80	A
		week 2	21 ± 2	19.2	0.26	A		53.6	1.39	A
		week 3	21.5 ± 2	19.2	0.26	A		53.3	1.19	A
		week 4	22 ± 2	19.5	0.41	A		52.9	1.10	A
		week 5	22.5 ± 2	21.1	0.35	A		55.4	1.23	A
		week 6	23 ± 2	21.2	0.22	A		53.7	1.39	A
		week 7	23.5 ± 2	21.9	0.45	A		55.4	0.99	A
		week 8	24 ± 2	22.3	0.29	A		55.2	1.17	A
		week 9	24.5 ± 2	22.7	0.41	A		54.6	0.57	A
		week 10	25 ± 2	23.8	0.36	A		55.1	1.17	A
		week 11	25.5 ± 2	24.4	0.26	A		57.1	1.98	A
		week 12	26 ± 2	24.4	0.19	A		57.7	1.71	A
	Transition period (autumn)	week 1	25.5 ± 2	24.7	0.17	A	52 ± 5	56.8	0.72	A
		week 2	25 ± 2	24.4	0.06	A		56.8	0.85	A
		week 3	24.5 ± 2	24.1	0.10	A		55.8	0.30	A
		week 4	24 ± 2	23.1	0.47	A		55.5	0.82	A
		week 5	23.5 ± 2	21.9	0.40	A		53.7	1.11	A
		week 6	23 ± 2	21.1	0.25	A		52.3	0.60	A
		week 7	22.5 ± 2	20.7	0.27	A		51.9	0.52	A
		week 8	22 ± 2	20.7	0.20	A		53.6	1.74	A
		week 9	21.5 ± 2	20.3	0.17	A		53.0	1.37	A
		week 10	21 ± 2	19.8	0.23	A		51.9	1.31	A
		week 11	20.5 ± 2	19.5	0.26	A		55.2	2.00	A
		week 12	20 ± 2	19.3	0.26	A		57.2	0.80	A

Table 4 Temperature/RH setting, average, standard deviation and tentative control level classification for the exhibition galleries (transition periods) (data 2010)

Exhibition gallery			Temperature (°C)				Relative humidity (%RH)			
			Setting	Average	SD	Classification	Setting	Average	SD	Classification
1	Transition period (spring)	week 1	22.5 ± 2	21.5	0.87	A	50 ± 10	46.5	3.95	B
		week 2	23 ± 2	21.4	1.05	B		47.9	5.49	C
		week 3	23.5 ± 2	21.4	0.83	A		49.1	2.42	A
		week 4	24 ± 2	22.8	1.11	B		46.7	3.72	B
		week 5	24.5 ± 2	24.7	0.61	A		51.9	3.40	B
		week 6	25 ± 2	23.4	0.59	A		48.5	3.90	B
		week 7	25.5 ± 2	24.6	0.59	A		52.7	3.52	B
		week 8	26 ± 2	23.6	0.14	A		54.3	1.96	A
	Transition period (autumn)	week 1	25.5 ± 2	26.3	0.53	A	50 ± 10	50.4	2.46	A
		week 2	25 ± 2	26.5	0.43	A		48.2	3.22	B
		week 3	24.5 ± 2	25.7	0.44	A		47.3	2.31	A
		week 4	24 ± 2	23.5	0.96	A		48.0	3.89	B
		week 5	23.5 ± 2	22.1	0.77	A		44.8	3.86	B
		week 6	23 ± 2	21.6	0.84	A		44.1	4.38	B
		week 7	22.5 ± 2	20.8	0.96	A		46.8	3.10	B
		week 8	22 ± 2	21.8	0.77	A		53.9	1.96	A
3	Transition period (spring)	week 1	22.5 ± 2	21.5	1.35	B	50 ± 10	55.2	4.10	B
		week 2	23 ± 2	21.4	0.94	A		55.8	3.42	B
		week 3	23.5 ± 2	21.6	1.11	B		56.2	3.97	B
		week 4	24 ± 2	20.1	1.16	B		58.8	4.78	B
		week 5	24.5 ± 2	20.0	0.87	A		55.6	2.60	B
		week 6	25 ± 2							
		week 7	25.5 ± 2	no data		no data				
		week 8	26 ± 2							
	Transition period (autumn)	week 1	25.5 ± 2	25.3	0.45	A	50 ± 10	54.1	2.99	B
		week 2	25 ± 2	25.5	0.42	A		51.7	3.26	B
		week 3	24.5 ± 2	24.7	0.43	A		50.4	2.33	A
		week 4	24 ± 2	22.3	1.10	B		51.8	4.37	B
		week 5	23.5 ± 2	20.6	0.88	A		49.5	3.80	B
		week 6	23 ± 2	20.0	0.84	A		49.0	4.74	B
		week 7	22.5 ± 2	19.7	1.26	B		51.5	3.84	B
		week 8	22 ± 2	20.8	0.88	A		57.8	1.62	A
5	Transition period (spring)	week 1	22.5 ± 2	20.0	0.96	A	50 ± 10	46.5	3.43	B
		week 2	23 ± 2	19.7	1.32	B		48.0	4.28	B
		week 3	23.5 ± 2	19.8	0.90	A		48.8	1.90	A
		week 4	24 ± 2	21.5	1.42	B		46.6	1.88	A
		week 5	24.5 ± 2	24.2	0.57	A		48.2	2.06	A
		week 6	25 ± 2	22.0	0.85	A		46.5	2.53	B

		week 7	25.5 ± 2	23.7	0.55	A			50.3	2.27	A
		week 8	26 ± 2	22.6	0.21	A			53.3	1.88	A
	Transition period (autumn)	week 1	25.5 ± 2	24.7	0.38	A	50 ± 10	52.2	2.08	A
		week 2	25 ± 2	25.2	0.43	A			49.2	2.83	B
		week 3	24.5 ± 2	24.3	0.36	A			47.5	1.57	A
		week 4	24 ± 2	21.7	1.19	B			49.7	3.74	B
		week 5	23.5 ± 2	20.1	0.79	A			48.4	3.53	B
		week 6	23 ± 2	19.6	0.85	A			47.6	4.78	B
		week 7	22.5 ± 2	18.6	0.99	A			48.4	3.68	B
		week 8	22 ± 2	19.4	0.78	A			56.1	1.68	A
7	Transition period (spring)	week 1	22.5 ± 2	20.2	0.80	A	50 ± 10	50.8	3.78	B
		week 2	23 ± 2	20.3	0.95	A			50.8	5.17	C
		week 3	23.5 ± 2	20.2	0.62	A			51.7	2.80	B
		week 4	24 ± 2	21.4	1.20	B			50.6	3.11	B
		week 5	24.5 ± 2	24.3	0.29	A			52.5	4.34	B
		week 6	25 ± 2	23.1	0.67	A			48.0	2.59	B
		week 7	25.5 ± 2	24.2	0.48	A			53.3	3.52	B
		week 8	26 ± 2	24.1	0.30	A			54.5	2.35	A
	Transition period (autumn)	week 1	25.5 ± 2	25.5	0.17	A	50 ± 10	51.3	3.07	B
		week 2	25 ± 2	25.3	0.31	A			51.3	3.58	B
		week 3	24.5 ± 2	24.6	0.28	A			49.5	2.64	B
		week 4	24 ± 2	22.5	1.02	B			49.7	3.99	B
		week 5	23.5 ± 2	20.5	0.83	A			49.1	3.15	B
		week 6	23 ± 2	19.9	0.55	A			48.8	4.51	B
		week 7	22.5 ± 2	20.1	1.16	B			50.6	4.26	B
		week 8	22 ± 2	21.1	0.76	A			55.7	1.86	A
8	Transition period (spring)	week 1	22.5 ± 2	20.6	0.89	A	50 ± 10	47.2	4.76	B
		week 2	23 ± 2	19.5	0.93	A			52.7	5.93	C
		week 3	23.5 ± 2	19.9	0.86	A			52.7	2.70	B
		week 4	24 ± 2	21.2	1.23	B			50.0	3.38	B
		week 5	24.5 ± 2	23.9	0.65	A			50.8	3.51	B
		week 6	25 ± 2	22.3	0.88	A			48.4	3.41	B
		week 7	25.5 ± 2	23.1	0.74	A			53.7	3.32	B
		week 8	26 ± 2	22.3	0.35	A			59.7	3.43	B
	Transition period (autumn)	week 1	25.5 ± 2	24.5	0.48	A	50 ± 10	57.1	2.83	B
		week 2	25 ± 2	24.3	0.71	A			53.5	3.63	B
		week 3	24.5 ± 2	23.8	0.64	A			50.9	2.91	B
		week 4	24 ± 2	21.6	0.98	A			54.0	4.70	B
		week 5	23.5 ± 2	20.4	0.87	A			52.6	4.20	B
		week 6	23 ± 2	19.8	0.87	A			52.2	6.89	C
		week 7	22.5 ± 2	18.9	1.02	B			52.4	4.22	B
		week 8	22 ± 2	19.7	0.83	A			58.7	1.55	A

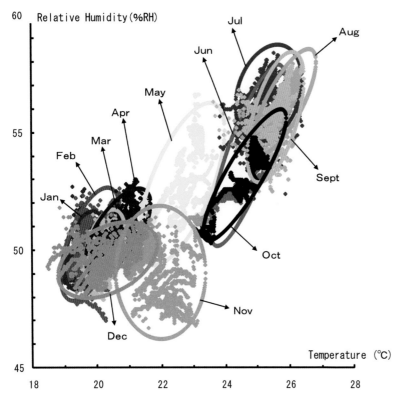

Fig 6
Monthly distribution of temperature/RH for general storage room 2 (data 2010).

results in a higher temperature fluctuation of ±5°C. In comparison to the special and general storage rooms, a greater fluctuation is observed in the galleries, especially in winter. Temperature fluctuation is within ±5°C in general and is often better during the transition periods while RH fluctuation is within ±10%RH in general and may be more than 10%RH in winter. These results show good correlation with the three levels of climate control adopted: the special storage rooms without seasonal adjustments are well controlled; the general storage rooms with seasonal temperature control are well controlled; and the exhibition galleries with seasonal temperature control are reasonably controlled.

A new attempt at analysis of the climate data is presented in Figure 6 using the data for 2010 from general storage room 2. An ellipse is drawn for each month to include 90 per cent of the data, with temperature on the *x*-axis and RH on the *y*-axis. The idea for the analysis comes from a previous study (Morita 1981). The distribution of the data permits a more detailed study of the transition periods. In spring, while temperature was changing gradually, RH changed and fluctuated in an irregular way, as illustrated by the two ellipses

for May and June, which cover approximately the same RH ranges. In autumn, however, temperature and RH changes occurred gradually, with the two ellipses for October and November showing less overlap both for temperature and RH.

Pest control measures

Pest control policy

In Japan, since the 1970s, a mixture containing methyl bromide has been commonly used as a fumigant for cultural properties. Following the Montreal Protocol on substances that deplete the ozone layer (which banned the production of methyl bromide in developed countries by the end of 2004), our pest control policy was revised to comply with the new regulations as follows:

For artefacts entering Japan for the first time, the use of gas fumigation is maintained, with ethylene oxide taking the place of methyl bromide. Chemical fumigation is performed to protect the Japanese ecosystem by preventing foreign pests from entering Japan through artefacts collected abroad. In 2007, a catalytic combustion unit for the detoxification of used ethylene oxide was introduced, in addition to a detoxification system through activated charcoal. At the same time, the existing fumigation chamber was improved by conversion into multi-function chamber that enables non-chemical methods to be used, namely, carbon dioxide treatment and anoxia treatment with nitrogen.

For artefacts that become infested in Japan, non-chemical methods are used. Carbon dioxide treatment is usually applied because it is easy to carry out within airtight enclosures. However, it is known that certain species of wood borer may withstand the treatment. In such a case, heat treatment may be the method of choice. Freezing is used for textiles, carpets and fur. In 2007, a new heating/freezing chamber was constructed at the museum to facilitate mass treatment.

Development of non-chemical control measures

In 2003, an Indian wooden boat measuring 18 metres in length, which is displayed in a gallery, showed signs of attack by wood borers. To remedy the situation, heat treatment was selected as it satisfied the following conditions:

- it was safe enough to carry out in a gallery
- it was applicable to large objects
- treatment can be performed within five days (excluding Saturday and Sunday).

An *in situ* heat treatment system composed of a thermal isolation container, heat generators and connecting pipes, which can be applied for the treatment of large objects, was developed by Morita, Sonoda and Hidaka (2004) (Fig 7). This system can be used anywhere where there is a power supply and can be easily moved. Hot air circulates inside the closed system, so the temperature and RH of the gallery are not affected. Prior to the treatment, the boat was hermetically sealed in an air barrier sheet and the internal air removed using a vacuum cleaner to reduce the risk of deformation due to water loss. Heat treatment of the boat was successfully performed within three days, including the time for setting up and dismantling the system. The temperature during the treatment was monitored using a wood log of similar diameter; the core temperature was maintained at over 55°C for 21 hours, thus largely satisfying the conditions necessary for effective heat treatment.

The heating/freezing chamber was first used in 2008, at the time of the renovation of the special storage room for carpets. Before the treatment, each carpet was condition checked, cleaned of dust, and photographed. All carpets were treated by freezing except for long ones which were treated with carbon dioxide in a huge airtight enclosure. The carpets were wrapped in a non-woven tissue, then in a PET film: 50 rolled carpets can be treated at one time in the chamber (Fig 8). The temperature inside the chamber decreased from the ambient temperature to –30°C in two and a half hours, remained around –35°C for five days, and increased gradually to room temperature after that. The temperature inside a control carpet remained below –35°C for more than four days and was below –30° for five and a half days.

Conclusion

We performed a systematic review of the museum environment data and evaluated the effectiveness of IPM preventive measures and the performance of climate control levels.

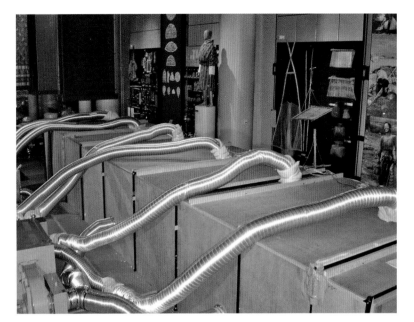

Fig 7
In situ *heat treatment of an Indian wooden boat in a gallery.*

A pest monitoring data analysis system was used to examine seasonal and geographic pest trends. Pest capture decreases from summer through to winter, and there are no clear seasonal characteristics in the species captured. From 1992 up to the present, the occurrence of pests has moved location from the storage and gallery zones to the other zone. In the storage zone, pest attacks tend to spread throughout the space while in the gallery zone pest attacks tend to occur on specific artefacts. The number of pests captured decreased on two occasions, in 2001 and in 2004, confirming the effectiveness of IPM activities: the former is the date from which we reinforced communication

Fig 8
Freezing of 50 carpets in a heating/freezing chamber.

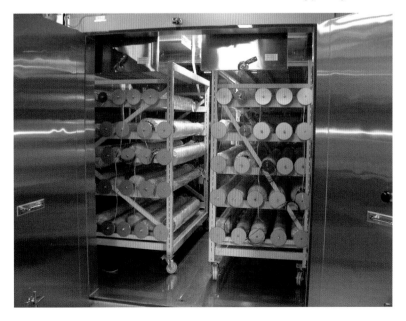

between different sections and the latter is the date from which we started to introduce a series of IPM activities.

The temperature/RH data analysis system aids understanding of the actual state of climate control levels. Generally speaking, temperature control is performed better than RH control and is better controlled in summer than in winter. The actual winter temperature averages are lower than the minimal temperature set point for many galleries and revealed to be more energy-efficient. A comparative study of standard deviation shows that the actual museum climate shows good correlation with the three levels of climate control adopted: the special storage rooms without seasonal adjustments are well controlled, the general storage rooms with seasonal temperature setback are well controlled, and the exhibition galleries with seasonal temperature setback are reasonably controlled.

At the National Museum of Ethnology, Japan, since 2005, the use of gas fumigation has been restricted to artefacts entering Japan for the first time. For the artefacts infested in Japan, non-chemical methods are adopted, including *in situ* heat treatment of large objects; pest control-related improvements at the museum include construction of a heating/freezing chamber and renovation of the existing fumigation chamber into a multi-purpose chamber.

Note: This paper described an example of the implementation of IPM and its re-evaluation in a Japanese museum. Interesting attempts have been made in other museums in the last decade, for example at the Kyushu National Museum where volunteers and NPOs are closely involved in museum IPM activities.

Acknowledgements

The authors wish to thank their co-workers at the National Museum of Ethnology, Japan, especially the staff from the Information Planning Section for their assistance. Invaluable research assistance was provided by Tomomi Wadaka, Yukako Kawamura and Sachi Hashimoto.

References

Morita, T 1981 'Microclimatic conditions within the National Museum of Ethnology'. *Bulletin of the National Museum of Ethnology* **6**(1), 159–82 (in Japanese with an abstract in English)

Morita, T, Sonoda, N and Hidaka, S 2004 'On-site heat treatment for large-size ethnographic objects'. *Bulletin of the National Museum of Ethnology* **28**(4), 539–70 (in Japanese with an abstract in English)

Sonoda, N and Hidaka, S 2008 'Between conservation and access: Implementation of integrated pest management at the National Museum of Ethnology, Osaka, Japan'. *Conservation and Access, Preprints of IIC 22nd Congress, London, 15–19 September 2008,* London: The International Institute for Conservation of Historic and Artistic Works, 88–92

20

To bag or not to bag? Treatment of a large Solomon Islands war canoe and the growing threat of drywood termites to collections in Australia

Colin Macgregor*, Heather Mackay† and Alex Roach‡

*Australian Museum, Sydney, Australia, colin.macgregor@austmus.gov.au
www.australianmuseum.net.au
†Australian Museum, Sydney, Australia, heather.mackay@austmus.gov.au
www.australianmuseum.net.au
‡Heritage Pest Management. Erina. New South Wales. Australia,
alroach@heritagepestmanagement.com (author for correspondence)

ABSTRACT

In March 2010 an infestation of drywood termites was found in a rare Solomon Islands canoe in the collection of the Australian Museum, Sydney. Drywood termites are among several pests classed as 'notifiable pests' in Australia, which means that the Australian Quarantine and Inspection Service (AQIS) must be contacted when an infestation is found or suspected. In consultation with AQIS it was decided that fumigation via low oxygen (anoxia) would be an acceptable means of eradicating the termites. However, the lack of environmental control in the off-site storage facility meant that low and fluctuating temperatures would be experienced, especially during winter months. To address this problem, an insulating tent was made with survival blankets around the canoe. Using three domestic oil heaters a minimum temperature of 20°C was maintained inside the tent for a treatment period of twelve weeks. The biology and habits of drywood termites are also discussed.

KEYWORDS

Drywood termites, low oxygen fumigation, anoxia, canoe, infestation

Introduction

The Australian Museum, Sydney has an active and comprehensive integrated pest management (IPM) programme that includes routine inspections of collection areas on site, as well as its off-site storage facilities. The inspections also extend to incoming loan and exhibition material, some of which are carried out in the presence of representatives from the Australian Quarantine and Inspection Service (AQIS). Australia has tight and rigid quarantine protocols for material imported from overseas and AQIS carry out inspections with the aim of preventing exotic pests and disease being brought into Australia. They also respond when exotic pest activity is found or suspected, which was the case when, during a routine inspection, insect frass was discovered on the floor near the stern of the museum's largest artefact, a rare wooden war canoe from the Solomon Islands.

This was immediately identified as being from drywood termite due to the distinctive size and shape of the pellets. Sheets of tissue paper were placed beneath the stern of the canoe in order to establish that the infestation was active, rather than old frass that had been

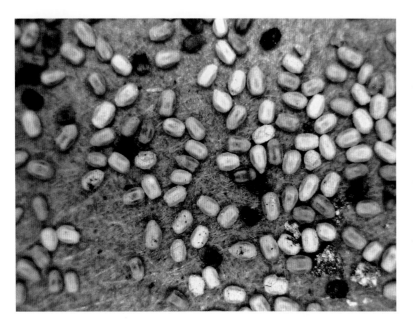

Fig 1
Digital microscope image of
drywood termite frass found
beneath hull of canoe.
(© Alex Roach)

of the Solomon Islands. It was commissioned in 1915 by Mr Harry Wickham, an Australian living in the Solomon Islands, and was used in a canoe race before being shipped to Australia where it was donated to the museum. It is constructed from planks of unidentified timbers around a long timber keel, and has high prow and stern posts elaborately decorated with shell inlays and carved figures.

It remained in storage for several years as it was thought impossible to display in the museum. In 1923, however, an ex-navy employee devised a method of hoisting the canoe through a window into the museum's Vernon Gallery (now the Indigenous Australians Gallery) and suspending it from the ceiling. The keel was reinforced with a steel support beam to spread the stresses on the hull. In 1983, after being on display for 60 years, the canoe was removed to make way for a new exhibition. It was stored behind display cases until 1994 when it was moved from the museum and taken to an off-site storage facility. It was relocated to another store in 1999, before being moved to the current storage facility in South Sydney in 2004.

Drywood termites

Drywood termites are so named as they do not require contact with the ground, like subterranean termites do, and instead they make colonies in dry pieces of timber, gaining moisture from the atmosphere and from the timber that they feed on (Fig 2). Colonies may be formed in dead trees or logs, but many species form nests in buildings within structural timbers, small wooden items and furniture. There can often be several colonies within a single piece of timber. Due to the small size of colonies they are easily transported, aiding their distribution to other areas (Hadlington *et al*

dislodged from movement of the wood. After two weeks a significant number of new frass had been ejected onto the paper (Fig 1). Examination of the frass by Australian Museum entomologists confirmed that it was from drywood termites. Further, the appearance of the frass was consistent with that of the West Indian drywood termite, *Cryptotermes brevis* (Walker). *Cryptotermes brevis* is regarded as the world's most destructive drywood termite and is a notifiable pest in Australia under the Diseases in Timber Act 1975 and, therefore, AQIS had to be notified.

AQIS officers responded by visiting the off-site store and carried out a thorough examination of the canoe, but were unable to make an identification of the particular species of termite due to the absence of adult specimens. Also, AQIS officers did not rule out the possibility of the infestation being caused by native drywood termites (Rickard 2010). The museum's DNA laboratory attempted to identify the termites by extracting DNA from the frass, however this was unsuccessful due to the small amount of insect DNA present in the pellets. Nevertheless, the presence of frass beneath the canoe was taken as a potentially serious infestation of a unique and irreplaceable collection item and a rapid response was deemed necessary.

Background

Fig 2
Drywood termites.
(© AQIS)

The canoe is a rare 14.5 metre long wooden war canoe or *tomako* from the Roviana region

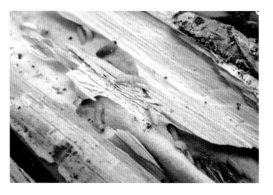

1985). They attack softwoods and low-density hardwood, and are found in many tropical countries throughout the world or coastal regions of temperate areas. Some species, such as the West Indian drywood termite, may also be encountered further inland (Horwood 2008).

Colonies spread through reproductive swarms or via sociotomy (that is, reproductives enter timber adjacent to the parent colony and develop independently) (Romoser *et al* 1998). Swarming reproductives (alates) fly toward light and enter suitable timber by tunnelling into wood through cracks or joins. When mature, drywood termite colonies are small with less than 10,000 individuals. Often the first sign of drywood termites is the appearance of frass on or under infested material. The frass consists of oval-shaped pellets with six raised ridges along the length of the pellet. Each pellet is approximately 0.5–1.0mm in length (Fig 3). Frass may be found in old galleries or tunnels, but is often ejected through a small hole measuring around 1–2mm. Alate wings (shed wings from reproductive termites) or damage to timber (for example, galleries in wood) may also be signs of activity (Gold *et al* 2005).

Treatment

AQIS usually require that objects suspected of termite infestation be treated through fumigation with chemicals such as methyl bromide. There are, however, significant health risks posed by such methods and the canoe could not be moved easily to a safe treatment site. Therefore, after discussions with AQIS, it was decided that the best option would be to treat the canoe *in situ* using low oxygen fumigation.

Low oxygen fumigation (anoxia) has been used in museums as a safe and effective method of treating insect infestations since the late 1980s (Gilberg 1989) and is routinely used for the treatment of infested artefacts, new acquisitions and loan returns at the Australian Museum. The treatments are carried out by encapsulating objects in high barrier film and using oxygen scavengers for the treatment of small objects, or humidified nitrogen for the treatment of large objects. In recent years, the museum has installed a nitrogen fumigation chamber which is used to perform most treatments.

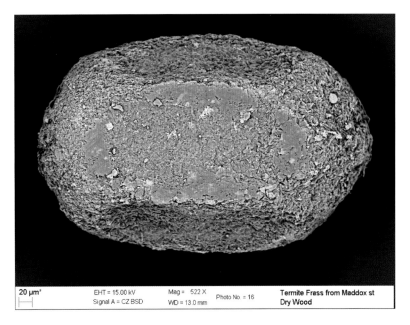

| 20 μm | EHT = 15.00 kV | Mag = 522 X | Photo No. = 16 | Termite Frass from Maddox st |
| | Signal A = CZ BSD | WD = 13.0 mm | | Dry Wood |

Whilst research has shown that anoxic treatment at 25°C will produce 100 per cent mortality in a matter of weeks, the heat source required to achieve this would incur considerable output and due consideration had to be given to the energy costs and safety of the collection material (that is the canoe). On numerous occasions at the museum low oxygen treatments of objects have been successfully carried out at 20°C when there are no time constraints on the treatment. Therefore, it was decided that it would be best to aim for a constant minimum temperature of 20°C and extend the treatment time as required.

Termites and low oxygen treatment

One consideration for the treatment of the canoe was the tolerance of drywood termite species to anoxic environments. Research into insect mortality rates conducted by Rust *et al* (1993) found that when adults and nymphs of the western drywood termite were exposed to <0.1 per cent oxygen at 25.5°C and 55 per cent relative humidity (RH) for a period of 96 hours, 100 per cent mortality was achieved. Reierson *et al* (1996) found that the western drywood termite was among the most tolerant species to an atmosphere deficient in oxygen, surviving for up to five days at <0.1 per cent oxygen at 26.5°C and 55%RH. Related research by Valentin *et al* (2002) showed results for furniture and large wooden objects infested with wood boring insects treated by low oxygen using the Veloxy® system. In order to kill all

Fig 3
Scanning electron microscope image of a single drywood termite frass, showing distinctive facets.
(© Australian Museum)

141

developmental stages of the insects 0.1–0.3 per cent oxygen was maintained for 15 days at 50–55%RH at 25°C in a 1m³ bubble. Further results showed objects treated at 20°C required a longer treatment time of 3–4 weeks, and success was dependant on RH levels and the characteristics of the wooden object undergoing treatment.

Difficulties of the site

A concern of treating the canoe at the off-site storage facility was the lack of environmental control. While there were no specific time constraints on the treatment, the off-site storage facility would be a difficult environment to treat the canoe in over the winter months due to low temperatures overnight of around 10°C and daytime temperatures around 15°C. Therefore, it was necessary to consider how the canoe could be insulated to maintain steady temperatures, thereby ensuring effective treatment. The canoe was to be treated in its current position, alongside a non-insulated brick wall and close to a large roller door. Various insulation materials were considered, such as re-using particle board shelving to construct a rigid enclosure around the canoe. However, this material was heavy and would require a degree of construction skill. An alternative material was emergency survival blankets. These blankets are used in emergencies where it is necessary to keep patients warm. They are composed of reflective aluminium-polyester film, are lightweight, inexpensive, available in

large quantities and, most importantly, are designed to insulate.

Another concern was ensuring that the canoe was structurally sound enough to be lifted and moved in order to allow for the canoe to be encapsulated in high barrier film. Termite tunnels and entry/exit holes were noted in wood close to the prow; the shells inlaid in putty along the hull of the canoe show signs of efflorescence, and the putty itself is brittle and prone to falling away from the wood. In general, though, the canoe appeared to be structurally sound. Fortuitously, the keel had been reinforced with a steel channel to support the canoe when it had been suspended from the ceiling of the museum. After the condition assessment was completed it was concluded that the wood did not show evidence of any major damage that could prevent it from being lifted.

Making the bag and temperature enclosure

High barrier film was used to encapsulate the canoe. A large tarpaulin was manufactured by heat sealing four eighteen-metre long panels together along the long axis (film width 1.3 metres). This resulted in a tarp measuring 18.0 × 5.2 metres. The tarp was rolled up and ready to be fed under the canoe.

Several staff members lifted the canoe onto custom-made dollies, each fitted with blocks of high-density polyethylene foam carved to support the canoe hull and prevent rolling (Fig 4). The film was then inserted under the canoe by elevating the canoe with the aid of a pallet jack inserted under each dolly. To prevent holing the plastic, a layer of thin polyethylene foam sheeting was inserted between areas where the canoe was in contact with the film and under the canoe where the film would be in contact with the dollies. The film was then folded over the canoe and the two edges sealed using hand and foot operated heat sealers. The canoe was purged with nitrogen humidified to 50 per cent. An oxygen level of 0.2 per cent was achieved after three days of flushing the bag with nitrogen (Fig 5).

A temperature enclosure was constructed around the canoe by joining a number of the survival blankets together. The blankets were taped to the ceiling, rear wall and floor to form

Fig 4
Blocks of high-density polyethylene foam used to support the canoe hull and prevent rolling on the custom-made dollies.
(© Australian Museum)

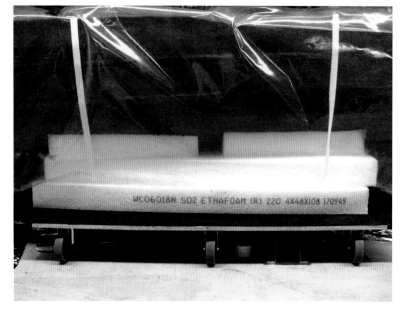

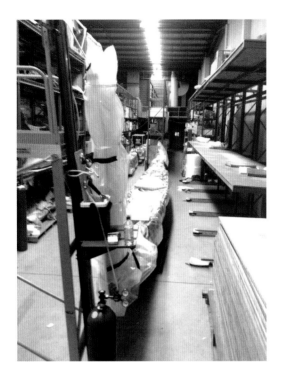

a three-sided tent before the canoe was rolled into place (Fig 6). A front flap was attached and an insulating 'sock' was fitted over the prow. In total, approximately fifty blankets were used to construct the tent (Fig 7).

Three domestic oil heaters were installed inside the tent, one at each end and one in a recessed alcove that had been prepared in the middle of the tent to ensure that the heater was kept a safe minimum distance from the canoe. Environmental data loggers with one metre long external probes were secured outside the tent at both ends and the probes were lowered inside. Initially, the heaters at each end were turned on and the conditions monitored over a one-week period to establish the baseline temperature. The output of the heaters was then increased until it rose above 20°C. To avoid the temperature inside the tent dropping below 20°C during the night, timers were also installed so that the heaters would switch on in the evening and turn off in the morning.

The canoe was encapsulated in the nitrogen atmosphere for a total of 32 weeks, until it was deemed that the treatment was complete. During the course of the treatment the oxygen level was regularly checked using a digital oxygen meter with a syringe tip to draw an air sample through the bag. At no time during the treatment did it rise above 0.2 per cent. Although the treatment commenced in June, it took until mid-October for the temperature

inside the tent to be constantly on or above 20°C. This temperature became achievable due to the warmer ambient conditions in spring combined with finding the ideal settings for the timers on the heaters to switch on at night and turn off mid-morning. The canoe was reinspected by AQIS officers in January 2011 and the treatment was considered to be a success. It was decided to keep the canoe encapsulated in the high barrier film as it would offer protection from further pest problems in the immediate future.

Environmental data

Environmental conditions from the data logger inside the treatment bag were downloaded and compared to the environmental conditions recorded in the main warehouse of the storage facility where the canoe is housed (Fig 8). For a twelve-week period from October 2010 to January 2011 when the treatment was complete, the temperature inside the bag did not drop below 20°C. Temperatures recorded ranged from 23.1°C minimum to 32.1°C maximum. This can be favourably compared to the warehouse temperatures recorded for the same period which ranged from 15.9°C minimum to 34.0°C maximum. Relative humidity levels recorded inside the bag for the same period ranged from 47.0 per cent to 64.77 per cent. Relative humidity levels recorded in the warehouse for the same period ranged from 24.7 per cent to 84.5 per cent.

Fig 5
The 14-metre-long Solomon Islands canoe encapsulated in a custom-made high barrier film bag with nitrogen flushing equipment connected to top of prow.
(© Alex Roach)

Fig 6
The canoe is rolled back into position in the incomplete temperature control tent.
(© Australian Museum)

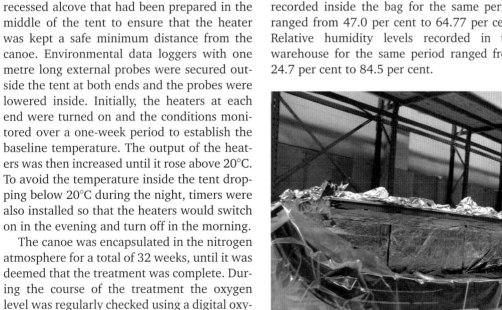

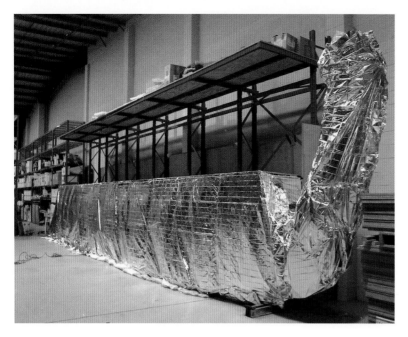

Fig 7
Approximately 50
emergency survival blankets
were used to create the tent
for the canoe.
(© Australian Museum)

the bag reached 30°C; the shortest period being seven hours and the longest period being 14 hours.

Conclusion

Drywood termites are among a number of insects regarded as exotic pests by the AQIS. Infestations can readily be introduced with artefacts or even building materials being brought into the area. These ongoing risks confirm the need for continual inspections and vigilance when introducing new materials to collection storage areas. The use of low oxygen environments has proved an effective method of eradicating possible termite infestations. However, stable temperatures need to be maintained to ensure insect mortality is achieved. In a treatment area where temperature control was problematic, the use of survival blankets to create an insulated treatment enclosure proved an effective means of elevating and stabilising temperatures within the space.

Significant results were also recorded during the warmer months where average minimum temperatures in the storage facility ranged from 14.4°C in November to 25.6°C in January. In these warmer months prolonged periods of elevated temperatures were recorded inside the bag. In November the temperature inside the bag did not fall below 25°C for a period of twenty-one days. During December and January there were 13 separate periods when the temperature inside

Acknowledgements

The authors would like to thank Ross Rickard, Michelle Ware and John Macdonald from Australian Quarantine and Inspection Service for information and support throughout the

Fig 8
Results of environmental
conditions recorded inside
the canoe bag compared to
external warehouse
conditions from October
2010 to January 2011.
(© Australian Museum)

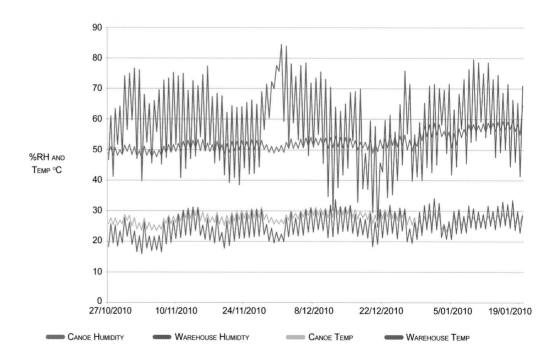

%RH AND
TEMP °C

CANOE HUMIDITY   WAREHOUSE HUMIDTY   CANOE TEMP   WAREHOUSE TEMP

treatment. The authors would like to also thank several staff from the Australian Museum including Michael Elliot for technical support, Anthropology staff for assistance with the canoe move and for background information on the canoe, and the DNA and Entomology departments for their efforts in identifying the termites.

References

Gilberg, M 1989 'Inert atmosphere fumigation of museum objects'. *Studies in Conservation* **34**(2), 80–4

Gilberg, M 1991 'The effects of low oxygen atmospheres on museum pests'. *Studies in Conservation* **36**(2), 93–8

Gold, R E, Glenn, G J, Howell Jr., H N and Brown, W 2005 *Drywood Termites*. E-publication E-366 (12/05). College of Agriculture and Life Sciences (Agrilife) Extension Programs, Texas A&M University System. http://insects.tamu.edu/extension/publications/epubs/e-366.cfm (accessed 1 September 2011)

Hadlington, P and Gerozisis, J 1985 *Urban Pest Control*, Sydney: New South Wales University Press, 217–32

Horwood, M 2008 'West Indian drywood termite'. *Primefacts – Prime Fact* 826. www.dpi.nsw.gov.au/__data/assets/pdf_file/0008/250487/West-Indian-drywood-termite.pdf (accessed 1 September 2011)

Maekawa, S and Elert, K 2003 'The use of oxygen-free environments in the control of museum insect pests'. Los Angeles: Getty Conservation Institute

Reierson, D A, Rust, M K, Kennedy, J M, Daniel, V and Maekawa, S 1996 'Enhancing the effectiveness of modified atmospheres to control insect pests in museums and similar sensitive areas' in Wildey K B (ed) *Proceedings of the 2nd International Conference on Urban Pests, Edinburgh, 7–10 July 1996,* http://icup.org.uk/reports%5CICUP742.pdf (accessed 21 April 2010)

Rickard, R, Australian Quarantine Inspection Service, pers comm, 4 March 2010

Romoser, W S and Stoffolano, JG 1998 *The Science of Entomology*, Dubuque, Iowa: WCB McGraw-Hill, 361–4

Rust, M K and Kennedy, J M 1993 *The Feasibility of Using Modified Atmospheres to Control Insect Pests in Museums*. GCI Scientific Program Report, Getty Conservation Institute, Marina del Rey, California, USA

Selwitz, C and Maekawa, S 1998 *Inert Gases in the Control of Museum Insect Pests*. Los Angeles: Getty Publications

Valentin, N 1993 'Comparative analysis of insect control by nitrogen, argon, and carbon dioxide in museum, archive and herbarium collections'. *International Biodeterioration and Biodegradation* **32** (4), 263–78

Valentin, N, Bergh, J, Ortega, R, Akerlund, M, Hallstrom, A and Jonsson, K 2002 'Evaluation of portable equipment for large-scale de-infestation in museum collections using a low-oxygen environment' *in* Vontobel, R (ed) *ICOM-CC 13th Triennial Meeting Preprints, Rio de Janeiro 22–27 September 2002,* Paris: International Council of Museums, 96–101

21

Webbing clothes moth in the Victoria and Albert Museum's British galleries: a successful campaign

Val Blyth* and Suzanne Smith†

**Conservation Department, Victoria and Albert Museum, London, UK,
valblyth@vam.ac.uk (author for correspondence)
†Furniture, Textiles and Fashion Department, Victoria and Albert Museum, London, UK,
s.smith@vam.ac.uk, www.vam.ac.uk*

ABSTRACT

This is the saga of how clothes moth activity in handling samples stored in a cupboard adjacent to galleries grew into a major infestation in the British Galleries of the Victoria and Albert Museum (V&A). The rise in moth numbers meant that a new strategy had to be devised to prevent the moths becoming established in the collection. One of the attractions of the British Galleries is the large number of textiles on open display, including tapestries, carpets, beds and other furnishings. All these were being placed at risk due to the rising number of moths as indicated by those found on the monitoring traps. The paper describes how the use of pheromone lures was used to pinpoint the source of the moth population beneath the floorboards of the Bromley-by-Bow room. After locating the source of the moth infestation in organic debris under the floor, ways of managing risks and eliminating the population were devised. Funding was secured to enable the lifting of the floorboards in the room so that the void underneath could be cleaned. The work had a tight schedule and a careful methodology had to be devised to keep the galleries open while the work was carried out.

KEYWORDS

V&A, textiles, webbing clothes moth, floorboards, voids, pheromone lures, IPM

Brief history of IPM at the V&A

The V&A was one of the first national museums in London to work with consultant entomologist David Pinniger to put in place an integrated pest management (IPM) strategy. Data collected from trapping and monitoring informed us of the presence and location of insect pests. It became clear that carpet beetles were established and breeding at the V&A site. Over the years, it was confirmed that the V&A had both Guernsey carpet beetle *Anthrenus sarnicus* and brown carpet beetle *Attagenus smirnovi*. We christened the latter vodka beetle. Treatment of infested objects and incoming objects at low temperatures became part of our preventive strategy (Smith and Blyth 2005).

Val Blyth and Suzanne Smith have worked together over the last six years refining a pest strategy for the British Galleries. Blunder traps are checked quarterly in the galleries by Visitor Services staff and moth lure traps are checked by the authors. Results are analysed and an annual summary report produced for the museum's Pest Management Group. Moths have only recently become a serious problem in the V&A and numbers have been documented using pheromone traps over the last five years.

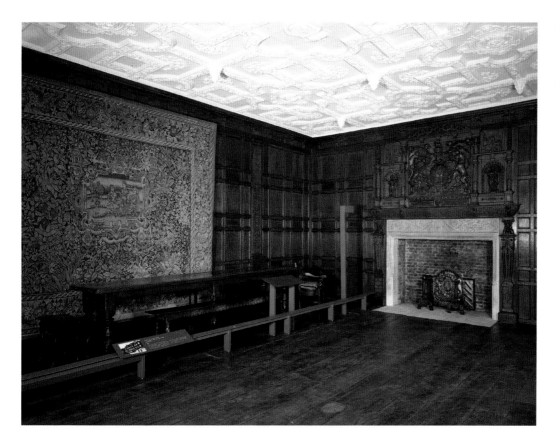

Fig 1
Bromley-by-Bow room,
British Galleries, V&A.
(Photo: V&A Photographic
Studio)

Moth infestation

Although webbing clothes moth *Tineola bissel-liella* is a very serious pest (Cox and Pinniger 2007), none were recorded in the V&A during the 15 years prior to 2005. In March 2006, moth-damaged wool samples were discovered in the British Galleries, in a store cupboard adjacent to the Great Bed of Ware and the Bromley-by-Bow room (Fig 1). The Bed of Ware is dressed with reproduction wool hangings, bedding and mattresses. The samples are provided to enable visitors to touch the different types of textile without touching the actual bed. The samples were immediately treated at a low temperature of minus 30° C for three days to eradicate all the live adults and larvae. We established that these objects were the original source of the moth infestation in the gallery. Further steps were taken to prevent the moths becoming established throughout the galleries, thus avoiding putting the large number of textiles on open display at risk.

In March 2007 AF Demi-Diamond™ moth lure traps (Fig 2) were used for the first time in the museum and these were checked monthly (Blyth *et al* 2008). These sticky boards impregnated with *Tineola bisselliella* pheromone were placed in a grid pattern. There was a steady increase in moths caught over the next few months. The grid of lures was extended to include an adjacent gallery, and subsequently there was a further increase in number of moths caught on traps. This increase prompted an inspection of all textiles on open display and, fortunately, none had become infested or damaged. Towards the end of 2007 the data collected revealed an even larger number of moths and so the number of traps was increased to cover a larger group of galleries. At this time, moth numbers were rising across London and south-east England (Pinniger 2009).

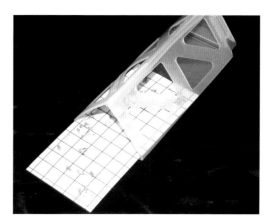

Fig 2
AF Demi-Diamond moth
lure.
(Photo: Valerie Blyth)

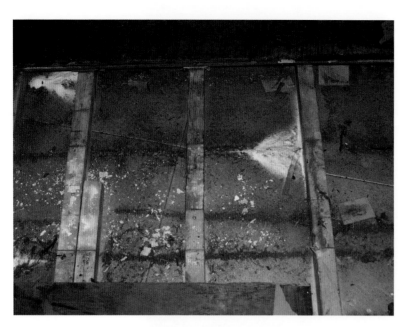

Fig 3
Organic debris under the floorboards.
(Photo: Valerie Blyth)

The Bromley-by-Bow room

Because of the large number of moths in the Bromley-by-Bow room, the tapestry, a carpet and two upholstered chairs were removed for Thermo Lignum™ heat treatment. Other smaller textiles were also removed and treated at low temperature. In response to the steady increase in moths caught, inspections were increased to fortnightly. The data collected allowed us to pinpoint the hot spot of moth activity to the Bromley-by-Bow room. Despite removing the textiles from the room, moth numbers continued to increase. In order to investigate a possible source it was decided to lift a portion of the floor and to establish whether the moths were living in organic debris under the floorboards. A small section of the floor was lifted to allow samples of debris to be collected. David Pinniger verified that the debris contained large numbers of cast skins of vodka beetle larvae, dead moths, moth pupae and head capsules of moth larvae.

Before the boards were replaced, a deep clean was carried out and Agrodust™, a desiccant dust, was applied in the gaps between the boards (Fig 3). There was a short period after cleaning when moth numbers fell, indicating that the targeted cleaning had been effective in reducing insect activity. But by the end of 2008 moth numbers had increased and it was clear that the limited action taken was inadequate to eradicate the problem.

The steady increase in the number of moths caught in the British Galleries during 2009 indicated that there was a real risk to the textiles on open display within these galleries. To remove this risk, we decided that the entire floor of the Bromley-by-Bow room should be lifted so that all the accumulated insects, dirt and debris could be removed. However, at this point there was insufficient funding to carry out this programme of work.

Inspection and surface cleaning of textiles

Because the funding was not available to lift the floor, we examined the textiles on open display to determine the risk of them becoming damaged. A report with a priority list of textiles was drawn up by Artemis Chaviara, who carried this out as part of her MA in Textile Conservation at the V&A. As a full-time student, Artemis was able to spend 280 hours over the summer of 2008 meticulously examining and cleaning the textiles (Blyth *et al* 2008).

Unfortunately, it was not possible to repeat this the following year. However, three students from the City & Guilds of London Art School Conservation course who had attended an in-house pest workshop taught by Val Blyth volunteered to come in to carry out the inspections and the surface cleaning of textiles on open display. The students also helped to check the moth lures in the British Galleries alongside museum staff. There were 32 moth traps in the British Galleries and moth numbers were recorded for analysis on a Microsoft Excel™ spreadsheet. During 2009, numbers were still high in the Tudor and Stuart galleries, and specifically in those traps around the Bromley-by-Bow room. This indicated that it was the probable source of the moth infestation.

Raising the floor

Following the work in the British Galleries, a report on trap analysis was used to justify the need for funding to lift the entire historic timbers in the Bromley-by-Bow room. Contingency funding was made available to carry out this work in February 2010. The company that originally laid the floor was contracted to carry out the remedial work. It was agreed that the timbers should be lifted, numbered and stored outside the room. The work was to be carried out over one week in March 2010. Visitor flow was redirected through another part of the gallery so that that the work be done without closing the galleries to visitors.

The planning for the works involved Conservation Department staff and curators from the Furniture, Textiles and Fashion Department. Visitor Services staff also had to be alerted to the work as any disruption would affect the gallery assistants as well as the visitors. Technicians removed the heavy items of furniture from the room prior to any work commencing. The textiles from the room had already been removed, with the exception of one very large glazed and framed textile.

After lifting and removing the floorboards, the debris was removed by vacuuming. Cut wooden fillets were added between the existing floor joists (Figs 4 and 5). Agrodust™ desiccant dust was applied in a grid pattern across the floor after the fillets were in place to inhibit further infestation. PUR Adhesive sealant was applied to the new wood before the historic timbers were nailed back in place. Mylands wax polish was used to fill the smaller cracks. The work was completed within the scheduled week.

The moth catches were closely monitored in the months following the work and it was very satisfactory to find that for the first month no

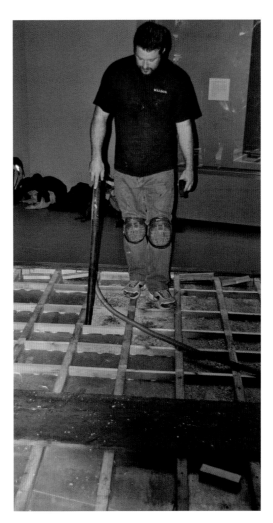

Fig 4
Vacuuming out the debris in the voids.
(Photo: Valerie Blyth)

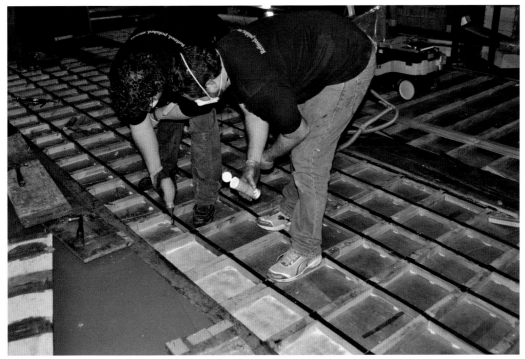

Fig 5
Application of desiccant dust and sealant.
(Photo: Valerie Blyth)

Fig 6
Total catch of moths on AF traps in rooms BG56–58 2008 to 2010.

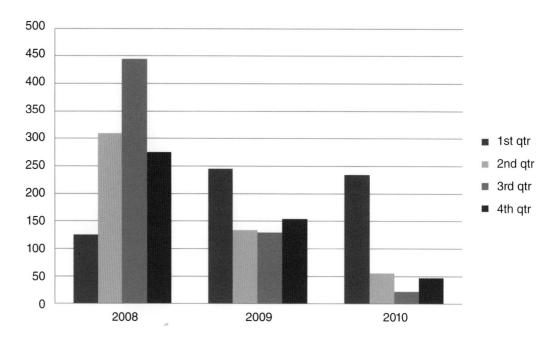

moths were caught in the gallery (Fig 6). Since then, moth numbers have remained at very low levels. As the risk from moth was now reduced *The Judgement of Paris* tapestry went back on display in late 2010.

Conclusion

This work has demonstrated the value of moth pheromone traps for pinpointing sources of webbing clothes moth infestation. Other points worth noting are:

- this is the first documented case of a *Tineola* population breeding in floor voids in a museum

- prior to treatment and at the height of the infestation, hundreds of moths were caught in the Bromley-by Bow room, but since the treatment, the numbers caught remain very low

- the treatment of the floor to eliminate the moth population has reduced the risk to all the textiles on open display in the British Galleries and it is now possible for the textiles removed from the Bromley-by-Bow room to be reinstated

- this project spanned six years and involved many people in the V&A. This underlines that for IPM to work successfully, there must be collaboration between all key staff.

Acknowledgements

The authors would like to thank: Sarah Medam, Nick Humphrey and Leela Meinertas, Furniture, Textiles and Fashion Department; Sandra Smith, Head of Conservation Department; Nigel Bamforth Furniture Conservation Department; and Marina Sokhan, Head of Conservation Course, City and Guilds of London Art School.

References

Blyth, V, Smith, S and Chaviera, A 2008 'Actions against the clothes moth at the Victoria and Albert Museum, British Galleries'. Poster presentation at ICON Care of Collections Group Workshop, *Pest Management in Practice*. Glasgow, 10 December 2008

Cox, P D and Pinniger, D B 2007 'Biology, behaviour and environmentally sustainable control of *Tineola bisselliella*'. *Journal of Stored Products Research* **43**(1), 2–32

Pinniger, D B 2009 'Clothes moths – numbers are surging'. *Pest Magazine* July/August 2009

Smith, S and Blyth, V 2005 'Prevention is better than cure'. *Victoria and Albert Museum Conservation Journal* **50**

22

Outsmarting the persistent moths: strategic planning for pest control at Nordiska Museet

Eva Helena Hakanen

Nordiska Museet, Stockholm, Sweden, eva.hakanen@nordiskamuseet.se
www.nordiskamuseet.se

Abstract

In the summer of 2009 an alarming presence of clothes moths was detected in one of the textile stores at the Nordiska Museet in Stockholm, Sweden. The store had a few years earlier experienced similar problems and undergone disinfestation. This time the moths were scattered throughout the room, and thus locating the source was impossible. It was essential to find a solution with a lasting result. A decision was taken to seal all the cabinets with plastic to prevent the moths from spreading and systematically to vacuum-clean and freeze all textiles. An extensive project was launched involving treating approximately 6,000 textiles. In the first phase, up to December 2010, 2,802 textiles were treated. It was established that 7.1 per cent of the textiles were infested with some form of remains of moth, mostly of older origin, and there were 24 objects which hosted living larvae. As a long-term solution all the cabinets will be covered with insect netting.

Keywords

Museum, textile stores, moth, disinfestation, pest control, freeze treatment

Introduction

The Nordiska Museet is Sweden's largest museum of cultural history, and has vast collections dating from the 17th century till today. These collections, accumulated from the 1870s to the present day, represent the life and work of the Swedish people and cover almost all types of material culture. A large proportion of the items comprise textiles and costume, many unique. One of the main textile stores holds over 25,000 textile items, costumes excluded.

During the summer of 2009 an alarming number of webbing clothes moth, *Tineola bisselliella*, were found in one of the main textile stores and the museum's most extensive disinfestation project was initiated. Up to December 2010, 2,802 rolled textiles had been vacuum-cleaned, frozen and re-positioned in the storage system.

This paper describes the issues and problems that have been faced and how the project has been planned and executed.

Several members of the museum staff have been involved with the project, the conservation section in an advisory capacity and the stores assistants on a practical level. One person has been working full-time for 15 months and one person part-time (3/5) for 11 months, with additional help from other members of staff.

The project has had valuable support from Anticimex, the pest control company, and Monika Åkerlund, curator in the Department of Entomology at the Swedish Museum of Natural History (www.nrm.se), in structuring the process and working out whether to use any insecticides inside the cabinets, and if so what type of insecticide.

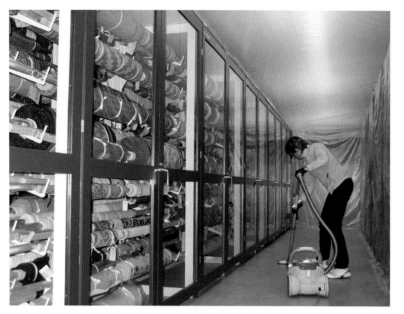

Fig 1
Rolled textiles in a cabinet,
the space being isolated
with plastic.
(Photographer: Eva
Hakanen; © Nordiska
Museum)

Background

The extent of the project was not obvious at first. It was difficult to understand the scale of the forthcoming work. However, during the planning process, a systematic and long-term strategy was developed. A few years earlier, the same store had previously had problems with moth and a large number of textiles were treated by vacuum cleaning and freezing. The problem had now reappeared, most probably due to the fact that items had been deposited in a non-controlled area before being returned to their rightful cabinets.

Within three large storerooms 25,000 textiles are stored, each about 400 square metres, and with a ceiling height of 3.4 metres. The infested store contains approximately 6,000 textiles, mainly of wool and linen. The room is furnished along the walls with cabinets containing drawers: in the centre there are 14

rows of tall cabinets, 6 per row, making a total of 84 cabinets. They contain bedcovers, tablecloths, carpets and curtains. The textiles in the cabinets are rolled on cardboard cylinders and hung on metal rods, on average 40 objects per cabinet. The largest textile measures approximately 1.60×2.0 metres (Fig 1). Wall cabinets with drawers include cushions with feather or down wadding, cushion covers, and wool embroidery.

The textile stores are subject to intensive pest monitoring on a regular basis. In the infested room a total of 111 Pre-Mal sticky traps have been placed on the floor, along the walls, and inside the cabinets during the period April to July 2009: 55 adult moths were captured in the sticky traps (Fig 2).

Planning stage

Preliminary research

The first meeting took place in September 2009. Prior to this a survey was carried out to establish the current situation. The statistics from the routine pest control inspection of 2008 and 2009 were analysed and a survey was carried out showing how the moths captured were distributed among the traps in the room. It showed that there was a scatter of insects throughout the stores. Studies of the available literature were carried out to update our knowledge on the life cycle and behaviour of the webbing clothes moth and information on the previous infestation was evaluated (Akerlund 1991). This attack had resulted in the vacuum cleaning and freezing of about 1,500 objects. As the work had extended over a couple of years, and the objects had not been satisfactorily isolated from each other, it appeared that the steps taken to remove the moths had probably not been successful.

Planning of treatment

One of the key issues in the planning stage was to estimate the level of working hours needed. Faced with having to take all the above facts into account, this proved to be a difficult task. It was agreed that a long-term solution was necessary. This would involve a complete review of all the textiles in the room in order, hopefully, to prevent a future outbreak.

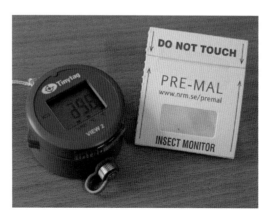

Fig 2
Data logger and sticky trap.
(Photographer: Eva
Hakanen; © Nordiska
Museum)

How, then, was this to be accomplished? The store room is large, and the big cabinets are not hermetically sealed. There is a gap of eight centimetres between the floor and the cabinet doors for the purposes of achieving adequate ventilation. This has obviously facilitated the infestation. Two further adjacent rooms, also containing storage for textiles, are subject to moth attacks, although to a lesser extent. It was estimated that a total inspection of the textiles would take approximately three years, and there was a severe risk that re-infestation would occur. A reasonably fast working process was therefore essential for a lasting solution.

Some issues that immediately emerged were how to:

- seal cabinets in order to prevent the spread of insects

- prevent the insects spreading from room to room

- adapt the work to accommodate access to objects for visitors, researchers and museum staff

- prevent the insects spreading while handling the objects

- decide in what order vacuum cleaning and freezing should occur

- determine the level of vacuum cleaning

- decide whether or not to treat the room and cabinets, and what pesticides should be used

- how to prevent or minimise future infestations.

Following discussions at the first meeting, the decision was taken that all textiles in the room would be vacuum-cleaned, sealed in plastic sheeting and frozen. All cabinets would be covered in plastic to isolate them and prevent the insects from spreading. Working areas and store rooms would be properly isolated from each other with plastic. Textiles were not to be moved unless protected by a cover. A working area would be organised close to the infested room where the textiles could be vacuum-cleaned and sealed in plastic before freezing. There was a concern that covering the cabinets with plastic film for a long period of time might create an undesirable microclimate inside them. As an insurance against adverse levels of relative humidity (RH), ventilation holes were cut in the plastic and these in turn were covered with Tyvek, a spun-bonded polyethylene

material allowing air circulation. Tinytag dataloggers were placed inside the cabinets and humidity and temperature were monitored (Fig 2). Measures were taken to create a closed system during the whole disinfestation process. A decision was taken to vacuum-clean prior to freezing. The textiles would then be protected in plastic until they were re-positioned in the cabinets.

Some compromises had to be made to enable access to some of the most frequently viewed objects while the project was ongoing. These textiles were vacuum-cleaned, frozen and placed elsewhere in the main store, away from the infested area. The remaining textile stores were closed to visitors, researchers and curators.

A discussion began on whether to treat the room and cabinets with a pesticide. Contact was made with the pest control company Anticimex and the Swedish Natural History Museum on what type to use.

The practical work process was then initiated. Ideas as to how to create adequate isolation between the rooms, how to cover the cabinets adequately with plastic and how to arrange the working area, took some working out. Luckily a member of the museum staff was able to provide simple but effective solutions.

Logistics and documentation

Once the broad outlines had been dealt with, the priority was logistics and documentation. It was essential to be able, if necessary, to find a specific object at any time while the project was ongoing. A large number of objects would be in circulation for a long period of time before being returned to their original location in the store (Fig 3). All textiles were therefore scrupulously labelled.

Due to lack of documentation from the previous moth disinfestation it was not possible to recreate the necessary information for future reference. To avoid the same situation occurring again, all stages in the work process are now fully documented. The objectives were twofold: firstly to be able to follow the movements of any object in terms of location, freezing date etc., and secondly to keep a record of treatment for every object.

Preparation

The necessary materials were purchased. All spaces, store rooms and cabinets were

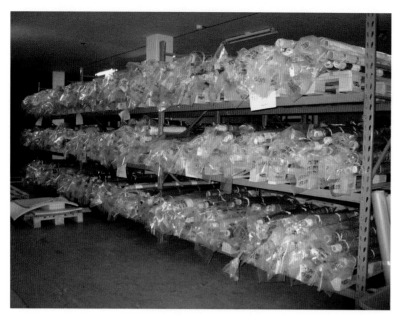

Fig 3
Treated textiles pending re-location in cabinet. (Photographer: Eva Hakanen; © Nordiska Museum)

Fig 4
Cabinets covered with plastic. (Photographer: Pamilla Håkansson; © Nordiska Museum)

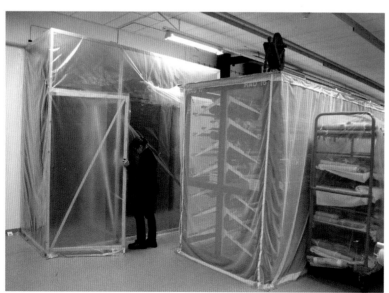

objects were removed at the same time and stored on a trolley. This was sealed with plastic and served as a temporary storage and quarantine centre until the objects were ready to be vacuum-cleaned.

The frequency of moth occurrence determined the order in which objects were to be treated, and which cabinet should come first. In accordance with the initial planning, the objects from the cabinets in the centre of the room, where the traps with most moths had been found, were the first to be vacuum-cleaned.

Practical work

In early November 2009 the vacuum cleaning started, utilising conventional vacuum cleaners, one cabinet at a time. The project also made it possible to introduce improvements to storage conditions. Old acidic tissue paper could be removed, and dirty cotton tape and infested gauze (a pharmaceutical product) could be replaced. In the cases where infestation was found, the gauze tubes around the rollers were replaced to minimise the risk of eggs surviving.

Textiles with remains of moth infestation were found relatively quickly. Fragments of excreta (frass), empty cocoons, and traces of mucus (webbing) were found. It was not possible to determine whether these were of old or recent origin. Many weeks of hard manual work went by with no sign of any living larvae or adult moth, although their existence was obvious through their frequency in the traps. In February 2010, after three months of vacuum cleaning, the first textile with living larvae was found. Finally! Paradoxically enough it was a relief, and acted as a confirmation for all the efforts made so far. An expensive and time-consuming project had been initiated. What an anticlimax it would have been if no living insects had been found. On the other hand, it was winter and not the main season for insect activity.

In March 2010 the textile with the highest number of living larvae, all 66 of them, was vacuum-cleaned (Fig 5). During the spring and summer some additional textiles were found to be infested with live moths. Once all the textiles with living larvae had been identified and treated, the number of insects found in the sticky traps decreased significantly.

meticulously isolated with plastic and the working area was made ready (Fig 4). An efficient workflow in terms of time was of the utmost necessity. Instead of folding plastic around every single object, plastic tubes that could fit around two rolled textiles were purchased and both ends were heat-sealed. Great care was taken to construct and arrange the working area so as to minimise strenuous positions and physical stress arising from repetitive work over a long period of time.

Initially, only the number of objects that could be treated in one day was removed from a cabinet. This strategy was soon to be revised, however, because the plastic proved difficult to open and re-seal. Instead, a large number of

A group of textiles infested in the previous outbreak, and isolated from the rest of the textile collection, was now vacuum-cleaned. Here, an active infestation was found. One bedcover housed 65 living larvae and, overall, out of 66 textiles, 44 had some form of clothes moth remains.

Most of the textiles treated were in fairly good condition. Relatively coarse textiles could therefore be vacuum-cleaned quite robustly. More fragile objects were vacuum-cleaned through a frame with nylon net.

Some 300 damask tablecloths were also stored in the room. However, no signs of infestation were present, nor could any insects be detected in any of the traps placed next to these textiles. The size and weight of most of the tablecloths made handling them extremely difficult; several table cloths being three metres long. The decision was made, therefore, not to vacuum-clean these objects, but only to freeze them.

Approximately one week before returning the textiles to their rightful places, the skirting boards, doorways, and empty cabinets were treated by Anticimex with the pesticide Demand CS.

The disinfestation process

As a typical example, the procedure for a bedcover will be described. The bedcover is removed from its place in the cabinet, and its object number and location is recorded. The textile is then placed on a trolley and taken to the working area, rolled out on the table and inspected, and thoroughly vacuum-cleaned on both sides. The object number on the tag fastened onto the roller and the number stitched on the actual object are compared. When necessary, old tissue paper and gauze is removed. Any traces of insects, such as excreta (frass), are easily detected since they fall onto the work table, and with a trained eye it is also easy to see other remains such as cocoons and traces of mucus. On textiles severely infested by larvae, the gauze has been eaten away (Fig 6).

When the bedcover has been vacuum-cleaned it is rolled up and placed in plastic tubing that is heat sealed at both ends. The plastic tube is clearly marked with the number of the object.

The treated textile is again placed on a trolley. When the trolley is sufficiently full it is wheeled into the freezer. Since the temperature does not reach below –27°C the rolls are

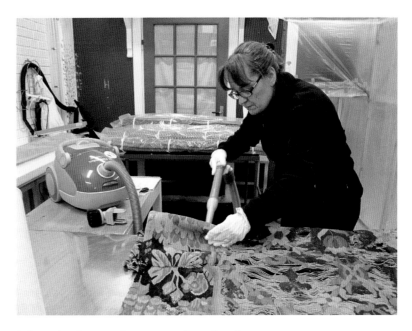

left in the freezer for two weeks. After freezing, all textiles are roughly sorted and placed on racks until they are returned to the cleaned and treated storage area. One cabinet at a time is filled with treated textiles. The area between the cabinets is isolated with plastic to minimise the risk of re-infestation (*see* Fig 1). The bedcover is then put back into its correct location and the object number is checked again.

The last stage in the process is to cover all the cabinets with a netting to prevent insects from getting inside. The netting is fastened with Velcro® in order to facilitate the opening of doors and the daily work in the store (Fig 7).

Fig 5
Work base, vacuum-cleaning textile with most live larvae, 66.
(Photographer: Pamilla Håkansson; © Nordiska Museum)

Fig 6
Infested bedcover rolled onto cardboard roller with gauze, cocoons and live larvae.
(Photographer: Eva Hakanen; © Nordiska Museum)

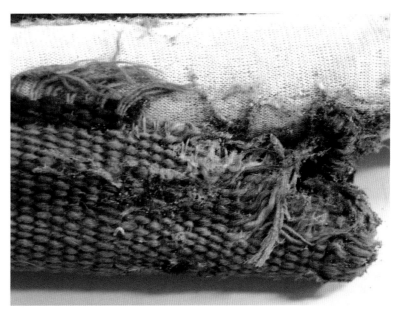

Fig 7
Cupboards with insect netting.
(Photographer: Pamilla Håkansson; © Nordiska Museum)

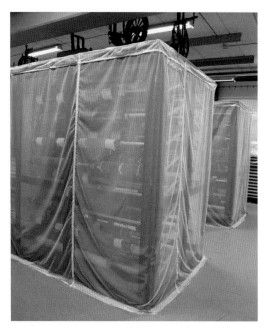

Results of the project

Summarising all the documentation in December 2010 resulted in the following information: 2,802 textiles had been treated and re-located; 24 objects were found to have held living larvae; a total of 200 objects (7.1 per cent) had various types of moth remains at different stages of severity; three cabinets among the 14 rows contained varying numbers of living larvae.

In summary, comparatively few textiles were actually infested. One might therefore ask whether all this hard work has been worth the effort. The answer is that it has been both important and necessary. By unrolling, inspecting, vacuum-cleaning and freezing the objects, and treating all the empty spaces with pesticide, we have a starting date and a base-line with no insect activity in the store. It is extremely difficult otherwise, in a large store room where a considerable number of moths appear in the traps, to identify the source. It was therefore essential to examine thoroughly, vacuum-clean and freeze all the textiles to gain control of the situation. As a knock-on effect we have updated our knowledge of the condition of all our objects. This has given us much better knowledge of how to react when the next outbreak occurs, hopefully in the distant future, and an understanding of the great necessity of rapid action. The members of staff have, above all, gained valuable knowledge and experience, and are now well prepared for any further strategic preventive work that may be necessary.

Conclusion

This project is far from complete, but we hope the end result will be successful and that all our efforts and measures taken have the effect of outsmarting the moths. In preserving the textile collection for posterity, a preventive strategy nowadays plays a naturally integrated part in the daily work of the stores, together with increased awareness of the behaviour of the moth.

Lessons learned include:

- the importance of covering all cabinets with insect netting to minimise the movements of any insects present
- minimising the movement of the textiles from one room to another
- using freezing as a routine precautionary measure after any movements of objects, for instance in connection with researchers, loans, and exhibitions
- establishing routines to keep units and rooms separate (keeping the doors closed)
- establishing routines for quarantine for infested objects by isolation, covering with plastic or netting
- when building or rebuilding storages facilities, the size of the storage area must not be too large. It is much easier to confine an outbreak with fewer objects involved.

It is estimated that this project will continue for three years before all 6,000 objects have been treated and returned to their locations in the store. All the cabinets will be covered with netting. This is a major task and, in consequence, work on the project has been at the expense of other museum activities.

The project has been both tedious and monotonous but still has had a few redeeming aspects, such as all the knowledge and experience accumulated along the way and, last but not least, all the wonderful textiles that we have had opportunity to inspect.

References

Åkerlund, M 1991 'Ängrar finns dom …? Om skadeinsekter i museer och magasin'. Stockholm: Museum of Natural History and the Swedish Museum Association

www.nrm.se (accessed 19 August 2011)

23

Quick, quick, put the lid back on!

Fiona Cahill* and Lucy Martin†

*Scott Polar Research Institute, Cambridge, UK, fc322@cam.ac.uk
†Scott Polar Research Institute, Cambridge, UK, lem28@cam.ac.uk
www.spri.cam.ac.uk/about/history/

ABSTRACT

Between 2007 and 2010 refurbishment of the museum and storage areas at the Scott Polar Research Institute (SPRI), University of Cambridge, highlighted ongoing problems with pests predominately webbing clothes moth (*Tineola bisselliella*) and vodka beetle (*Attegenus smirnovi*). Limited integrated pest management (IPM) and sporadic collections management resulted in a significant number of objects (particularly textiles) being damaged. Improvements in storage and access to walk-in cold rooms created the opportunity to address this problem and limit further damage to the collections.

KEYWORDS

Integrated pest management, IPM, textiles, freezing

Historical background

The Scott Polar Research Institute (SPRI) was founded in 1920 as a memorial to Captain Scott and his companions who reached the South Pole in 1912. The intention was to create a place where information about the Arctic and the Antarctic could be collected and stored and where scientists and polar explorers could meet for the furtherance of knowledge (Debenham 1945). It is now a major international centre for polar research, information and expertise. The building in which the Institute is housed consists of a variety of different architectural spaces. The original building was built in the 1930s; a major extension added in the 1960s now forms the largest part of the Institute. Further works were completed in the 1980s and the Rotunda, built as an extension to the Library, opened in 1998. The basement in the 1968 building contains the museum stores, the map room, compact shelving for library storage, two cold rooms and a field store and workshop. The ground floor comprises the museum, which is open to the public and is sited mainly within the original 1934 building, opening onto the main foyer in the 1960s extension which houses the special exhibitions gallery, lecture theatre, administrative offices and laboratories. The first floor is mainly given over to the Library, which extends through both the old and the new buildings. The Friends Room also located on this floor, which provides a general working area and access to the Rotunda. The environmentally controlled archives and photographic stores are housed on this floor. On the second floor are further offices for staff and researchers, alongside another large library stack space and workspaces for students. The old part of the building is also accessible from this floor and contains a meeting room, office and more library storage in what was once a gallery space. Access to the attic is also from this floor. The Rotunda is situated to the side of the building and is accessible from all floors. It contains bookshelves and individual study spaces.

Overview of the collection

The SPRI's collection comprises over one million items, including 2,116 archival fonds (which may contain large numbers of individual objects), approximately 100,000 photographic items, including film, and over 5,000 artefacts. All of these objects are susceptible to damage, whether it comes from pests or the environment or, in most cases, both.

Initially, objects were deposited at SPRI for storage and research rather than donated. Literature in the archives tells us that, although these objects were seen as part of a museum collection, they were not necessarily displayed in an archetypal museum fashion. With almost no funds for museum staff, little change was made to the museum displays between 1980 and 2006.

Historically, reserve collections at SPRI were stored throughout the building in any available space, with the majority being stored in a basement room now identified as B5. The larger, oversize objects such as kayaks and sledges have been stored in the attic space. A number of building programmes over the years since 1934 have changed the configuration of the building. Access to the attic was once fairly straightforward, but it is now an extremely inconvenient place to get to. Other items have been stored in filing cabinets in different locations, in the basement, in cupboards under stairs, in staff offices and on display in public areas. The staffing levels of both the museum and the archive consisted, for many years, of one part-time paid employee assisted by varying numbers of volunteers. With a subject matter that attracts constant attention, many of the collections' management issues were sidelined in favour of dealing with enquiries.

The artefacts in the collection are constructed from a variety of materials including, amongst others, textiles, leather, metal, ivory, wood, fur and plastic. The cultural range to which these artefacts relate encompasses both Polar Regions and represents a broad spectrum of activity from the Heroic Age of Antarctic exploration, Inuit and northern cultural materials to present-day science and research projects. The textiles in the collection not only include clothing, but also a variety of non-costume-related objects such as ski bindings, whips, man hauling harnesses, reindeer harnesses and dolls. The most vulnerable materials in the collection in regard to pests were the textile and wooden items.

A number of objects were displayed in the Institute's publicly accessible galleries from 1934 onwards. These galleries were along the lines of a cabinet of curiosities popular at the time, then used as a teaching or reference collection, as opposed to what we would expect from a museum display today. From 1947, when the museum re-opened after the Second World War, the objects on display were not always sympathetically mounted and the case materials used had the potential to cause damage to the objects (Fig 1).

However, these objects were to a certain extent protected from the pest issues occurring in the storage areas. It is interesting to note that when these objects were taken off display during the museum redevelopment, the textiles usually had a box of crushed mothballs in the base of the case. This suggests that they had come from an active moth environment within the building, namely the stores in the basement and that moth had already been identified as a problem.

Identifying key problem areas within SPRI

The redevelopment of the museum and stores offered the perfect opportunity to assess the condition of the collection and to address the pest problem within the stores. The primary store in the basement, B5, was a room approximately 5.5m × 4m × 2.7m containing metal racking and shelving. It held approximately 2,500 objects of which 711 were textiles, the majority of which were hung tightly packed on two clothes rails. The rest of the textiles, predominately mittens, socks and scarves were packed away in boxes on the shelves (Fig 2).

In 2002, in an attempt to alleviate the overcrowding in B5, some of the hanging textiles were moved into a room in the basement room

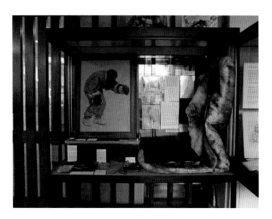

Fig 1
Display case in old SPRI museum pre-refurbishment.
(Photo: SPRI/Fiona Cahill)

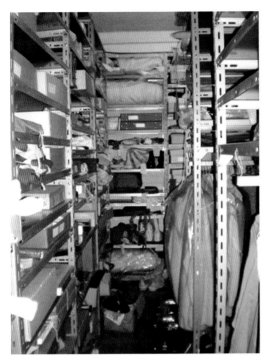

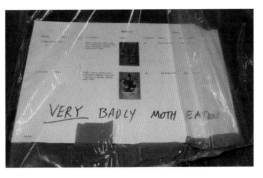

Fig 2 (far left)
B5 store pre-refurbishment.
(Photo: SPRI/Fiona Cahill)

Fig 3 (left)
Indication of an existing
moth problem.
(Photo: SPRI/Fiona Cahill)

identified as B11 and subsequently called the Textile Store.

A freezing programme was initiated for these textiles but it unfortunately did not work. Due to lack of space, frozen textiles were put back into the same room as untreated textiles, resulting in recontamination. Neither B5 nor B11 had ventilation or consistent environmental and pest control within the room. In all stores a number of empty tins of insect repellent were found. During David Pinniger's initial visit in 2007, large numbers of moth in the Textile Store were noted, both in the boxes and in the carpet (Pinniger 2008). The carpet tiles were removed in 2008 (Fig 3).

As SPRI is principally a research facility, staffing had been focused more on the research aspect rather than on collections care. Along with other responsibilities, members of staff would sporadically carry out housekeeping in the store areas, consisting mainly of infrequent vacuuming with a domestic vacuum cleaner. Due to lack of training and inappropriate equipment this vacuuming may have led to the further damage of already vulnerable items. The pest problems were not limited to the storage areas. Members of staff eat at their desks with food debris accumulating around work areas due to no dedicated staff room. Food wrappers and fruit peelings were also put into waste paper baskets, which although emptied on a regular basis, provided two ready food sources for pests. It was not always possible for the cleaners to vacuum the office areas regularly as access was limited.

For a number of years there has been a wasps' nest at SPRI, probably in the eaves, although the location is still unknown. However, the large quantity of dead wasps found in the attic and on windowsills in the upper floors suggest the nest is somewhere in the roof space. Inadequate housekeeping of dead spaces, especially in the library, resulted in debris collecting under the stacks.

Solutions

The initial IPM system at SPRI was set up by Bob Smith, Museum Redevelopment Project Manager, in 2007. He placed pest blunder traps around the building to assess the extent of the problem. In the stores, these showed that the textiles were the only evidence required to flag up a moth infestation (Fig 4).

Bob Smith was also responsible for inviting David Pinniger to the Institute in December 2007 to conduct an assessment to determine the scale of the problem. In his report following the visit David highlighted the 'serious problem with webbing clothes moth in the store' that was 'exacerbated by the poor and congested storage conditions in the basement and attic' (Pinniger December 2007). David also discovered three *Trogoderma angustum* in

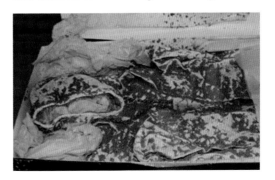

Fig 4
Fur parka (Y.2002.5.3a)
totally eaten by moth.
(Photo: SPRI/Fiona Cahill)

the library. This find was confusing at first as they commonly attack plant and animal specimens, in particular herbariums. None of these foodstuffs were present in the library, and only one on record in the store. Subsequent investigation determined that these pests were actually feeding on dead wasps. A four-page guidance document prepared by David in January 2008 contained many recommendations. These recommendations stated primarily to freeze the textiles and move them to a clean storage area and to implement a full IPM programme (Pinniger 2008). David would revisit the Institute in December 2009 and July 2010.

Building on the system set up by Bob Smith, it was decided to dedicate two members of the collection staff to be responsible for monitoring and maintaining the IPM system: the conservator (employed on a temporary contract as part of the museum refurbishment) and the Picture Library Manager. They both attended the IPM 1 and 2 courses run by the SHARE (MLA Renaissance scheme) taught by David Pinniger.

As part of the redevelopment a dedicated textile store, Museum Store B (MSB) was incorporated into the new main store Museum Store A (MSA). The Textile Store (B11) was earmarked as space for a conservation workshop, so the priority was to re-freeze the boxed textiles and move them into MSB.

Historically, insects have been eradicated by fumigating with chemicals such as methyl bromide, phosphine and ethylene oxide. Moths have been eradicated with chemicals such as naphthalene and paradichlorobenzene (found in moth balls). Although these treatments were very successful, the use of such chemicals has been restricted over time due to health and safety or environmental considerations, leaving few options open to museums. Chemical solutions have been replaced with treatments including anoxia (removal of oxygen), heat treatment (raising temperatures to over 50°C) and freezing (Pinniger 2003). Some of these treatments require specialist equipment. In its capacity as a research facility SPRI is very fortunate to have two walk-in commercial freezers, known as the cold rooms. The first cold room is set at −30°C and the second +4°C. Following the literature (Canadian Conservation Institute 1997; Pinniger 2003) the textiles were wrapped in acid-free tissue paper and then in plastic sheeting with the excess air

expelled. Although it is recommended that textiles are only left in the colder of the two freezers for three days it was more expedient to leave them for seven days. This allowed the textiles to be moved on the same day each week. With a large volume of textiles and a small number of staff it worked very well to have a specific time set aside each week. After seven days the textiles were moved into the second cold room to be brought gradually up to temperature, minimising problems with condensation. Once the textiles were removed from the second cold room they were left on a trolley in the store for a few days to acclimatise before unwrapping. A systematic approach was taken to freezing the textiles as the locations had to be changed on the museum database. Two people were needed to move the boxes between the freezers and into the new store as some of the textile boxes were very large and unwieldy. Eventually 1,307 textiles went through the two cold rooms over a four-month period. B11 was then refurbished as a conservation workshop (Fig 5).

During this time it was noted that although the MSA and MSB storage areas were completely new and empty of objects, vodka beetle larvae were still being trapped in high numbers. It was thought that they might be coming from the plant room adjacent to the new store, so this area was also refurbished and the floor filled and levelled. The room also contained a large volume of ducting that formed part of the air conditioning unit for the lecture theatre above. During David Pinniger's visit to the Institute in December 2009 he noted that although the new stores were very clean the source of the high number of vodka beetles larvae caught on these traps was unclear. On his advice regular housekeeping was complemented with Constrain (permethrin-based insecticide) and Agrodust (silicon dioxide

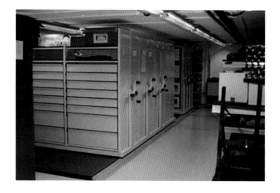

Fig 5
SPRI Museum Store A.
(Photo: SPRI/Fiona Cahill)

desiccant dust) in the store areas (Pinniger July 2010). It was a disappointment that these areas could not be made completely free of pests before objects were relocated from the old storage areas, but the precautions in place were felt to be sufficient to risk moving into the new store.

A number of other procedures were introduced as part of the IPM programme. The conservator talked through the 'Insect Pest found in Historic Houses and Museums' poster (English Heritage/Collections Trust 2009) and the purpose of the blunder traps with the cleaning staff. This approach was very successful, as the cleaners understood why they were being asked to clean in dead spaces, radiator grills, and so forth, and so the levels of housekeeping improved (this was noticeable as dust bunnies were no longer a main feature on the blunder traps). It did become clear that the one vacuum set aside for library and archive use was old and not maintained regularly. To assist in regular cleaning and reduce cross-contamination two new vacuums cleaners with HEPA filters were purchased for the sole use in the archival stores and the new museum store.

None of the objects are on open display in the new Polar Museum. This greatly facilitates IPM as the traps (pheromone and blunder) can be inserted into the space in the base of the cases. If a problem arises, it is contained within the case. It is also easy to visually inspect the cases.

As the majority of staff did not have a museum or collections background the conservator wrote a short procedural document about bringing potentially problematic materials into the Institute. It outlined how objects and textiles should not be taken straight to staff offices, but should go through quarantine and freezing procedures first. An email was sent to all staff alerting them that the electronic document was available on a shared drive. Subsequently, a number of personal items belonging to Institute staff have gone through the freezers, whereas in the past this would have been unlikely as they were unaware of the risk. A budget has been set aside for IPM materials and the conservator has the responsibility of ordering and maintaining these supplies.

At first, our main priority was to address the issues in the storage areas. However, as with most things, this snowballed beyond the stores

and into other areas of the Institute. Blunder traps in the bibliographers' office within the library on the first floor captured high numbers of vodka beetle larvae (Fig 6). It transpired that the carpet had been in place since the extension was built in 1985. With the volumes of books passing through this area and inadequate space, access for cleaning soon became problematic. As mentioned previously, staff members eat at their desks and all these factors combined to establish a vodka beetle larvae hot spot. After consultation with the librarian, the bibliographers and Institute administrator the office was cleared out refurbished and a new linoleum floor was laid. This also gave the bibliographers the opportunity to de-clutter, resulting in better access for the housekeeping staff. Blunder traps in this area are now empty and vodka beetle larvae have been eradicated. However, as one problem is solved another rears it head. In the Old Library itself a number of adult beetles were found on traps and it was discovered that the heating system in some areas was boxed in and had not been cleaned for some considerable time, allowing an accumulation of dust and debris, a perfect breeding ground for pests.

Unfortunately, it was not possible for the cleaning staff to access these areas as it required a maintenance person from Estate Management Building Services (EMBS) to remove the grilles, which they were advised not to do as asbestos, which should not be disturbed, had been found in parts of the system. Although Constrain could be used temporarily to deal with pests currently present, it would be preferable to clean out the grille and remove the food source for future pests.

Conclusion

Since the IPM programme was instituted, there has been a significant drop in the number of moths caught in traps, with only three trapped over a 15-month period. In the storage areas this was due to the introduction of IPM, new racking, good housekeeping and a strict freezing policy. Within the wider Institute the number of pests has also fallen, due primarily to the improved housekeeping standards. Data collection of pest numbers has also been a good tool for prompting overdue works, for example, replacement flooring in the bibliographers' office.

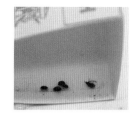

Fig 6
Adult vodka beetle found on SPRI Old Library blunder trap.
(Photo: SPRI/Fiona Cahill)

With few resources and staff this project was initially time and personnel intensive. As there were other calls on time, discipline was required to maintain a regular schedule for trap checking. A free and effective way of reducing pests is to inform staff about the reasons for the programme. IPM needs to be monitored regularly, and this is an ongoing process at SPRI as there are still pest problems to address. As staff members are employed predominantly on short-term or temporary contracts, it is important that the system is simple so that, if necessary, it can be taken over quickly by new personnel. Ultimately, the project aimed to reduce the threat to the collection from insect pests and this has been successfully achieved. David Pinniger's visits are now made annually, during the summer at the height of pest activity and any problems should be evident then. Stores are now vacuum-cleaned every two weeks and deep cleaned (for example on the top of racking) every six months.

Unfortunately, in the SPRI building it has not possible to eradicate vodka beetle completely but they can be controlled. The next stage is to thoroughly investigate the roof space, once the large objects have been moved, in order to locate the wasps' nest. Looking beyond the museum galleries into the Institute, problems still need to be addressed in areas such as the basement map room, which may harbour pests. The ultimate aim is to be able to carry out a trap inspection that only takes one morning every three months!

References

Canadian Conservation Institute (CCI) 1997 'Controlling insect pests with low temperature'. *CCI Notes* **3**. Ottawa: CCI www.cci-icc.gc.ca/publications/ccinotes/enotes-pdf/3-3_e.pdf (accessed 1 September 2011)

Debenham, F 1945 'Retrospect: The Scott Polar Research Institute, 1920–45'. *Polar Record* **4**(29), 223–35

English Heritage 2009 'Insect Pest found in Historic Houses and Museums'. Poster. London: English Heritage and Collections Trust

Pinniger, D 2003 'Museum artefacts, saving our treasure – controlling museum pests with temperature extremes'. *Pesticide Outlook, Journal of the Royal Society of Chemistry* **1**,10–11

Pinniger, D B 2008 *Pest Management – A Practical Guide.* Cambridge: Collections Trust

Pinniger, D B (Consultant Entomologist). SPRI site visit (internal reports) December 2007, January 2008, December 2009 and July 2010

Pinniger, D and Winsor, P 2004 *Pest Management: A Guide for Museums, Libraries and Archives.* London: Museums, Libraries and Archives Council

SHARE – A network for sharing skill. MLA Renaissance East of England. www.mla.gov.uk/what/programmes/renaissance/regions/east_of_england/info_for_sector/SHARE (accessed 1 September 2011)

24

But how do I know I've got pests?

Jane Thompson Webb

Birmingham Museums and Art Gallery, Birmingham, UK,
Jane_Thompson-Webb@birmingham.gov.uk

ABSTRACT

The concept of integrated pest management (IPM) has permeated large parts of the museum sector and its practice is now commonplace in the larger museums. The United Kingdom's Museum Accreditation Standard includes a requirement that museums inspect for insect infestation and identifies the link between it and housekeeping. It would seem, therefore, that IPM is now fully integrated into day-to-day practice.

But for many museums in the independent or voluntary sector this is not the case, because they have no one on the staff who has received training in integrated pest management (IPM) and information in this area can be difficult to access. This can put collections at risk from pest infestation. In the West Midlands region, the Renaissance At Work (RAW) programme has sought to bridge this gap. Through innovative practical training and encouragement, museums of all sizes are given the knowledge and skills to implement IPM. Some museums grasp the concepts of pest management more successfully than others, and there are definite success stories, illustrated by the case studies of the Cider Museum and the Shakespeare's Birthplace Trust.

Even after training, implementing IPM can be problematic when it is only a small part of a person's job. To help museums post-training and to support those who have not been able to attend courses, RAW has developed a CD and web resource to help identify insects and solve pest-related problems. The resource also includes an insect pest recording database, which will give us, for the first time, an overview of what pests are present in the region. It is intended that this will become a tool for pest recording across the UK.

KEYWORDS

Integrated pest management, IPM, training, advice, Renaissance programme, pest CD, 'What's Eating Your Collections?'

Introduction

'It is a truth universally acknowledged that a museum in possession of a collection must be implementing an integrated pest management (IPM) strategy' (after Austen 1813). For those of us working in large institutions with collection care, conservation or collections management staff, this is quite likely to be the case. We have staff who have undergone professional training, attend refresher courses and have tasks related to pest management allocated. We can sleep soundly knowing that there is little risk of any insect eating our collections without us knowing about it.

But what of the smaller museums with few, if any, paid staff? In these institutions, it is common for the staff to fill multiple roles, and whilst stewarding, building security and perhaps documentation are obvious and necessary tasks, IPM is not. Except, of course, it is we, the fortunate group that knows about such things, that recognise the importance of an IPM programme. It helps us to protect our buildings and the collections they hold by spotting infestations before they become too serious. It also helps to show us where we have environmental problems, which might have passed unnoticed but for the presence of certain insects. In addition, it helps us to identify objects at particular risk from pests so that we can use our time efficiently when checking for pests.

How are smaller museums to find information about IPM to enable them to prevent their collections from being eaten away? They could pay to attend a pest course, but the problem is that smaller museums are often operating on a very limited budget and cannot justify the expense of attending a commercial course. If we accept the concept that museums exist as a community protecting the heritage of the nation, then we need to find a mechanism whereby smaller museums can access information about IPM and that they can do this in an affordable way. There also needs to be a mechanism to support staff trained in IPM to help them to build their confidence and ensure that they do not feel they are working in isolation.

This is a fine aspiration, but how can such training and support be provided? In the West Midlands region we have found the answer, and the answer is RAW. RAW stands for Renaissance at Work, a deliberate play on words to show that the Renaissance in the Regions programme for England can be useful to museums other than the large 'Hub' museums, and that it can provide help and support to individuals working in them.

Training was once provided by the Area Museum Services/Councils and information on a range of collection care issues, including IPM, was available through the books published by the Museums and Galleries Commission (later Museums, Libraries and Archives Council (MLA)). With the demise of the regional bodies and the more strategic stance adopted by MLA, there was no training provision for museums and a knowledge gap began to be evident. One of the aims of Renaissance in the Regions programme was to bridge that gap, identified by Porter (2002), by encouraging the Hub museums to be sources of information and training for the museums in their region.

Birmingham Museum and Art Gallery is the lead museum in the West Midlands Hub. As part of the original business plan, put together in 2004, the museum decided to offer a training programme covering all aspects of collection care. A programme of 42 courses covering 15 subject areas was devised and training began in October 2004. I was appointed into a Renaissance-funded post with a dual role: part of my time was spent working on collection care for Birmingham's museums, and part of my time was spent co-ordinating and leading the RAW programme. This is a key point; it was my job to run courses, and we were obligated to report to MLA the number of courses run as part of the Renaissance reporting process. For any training programme to be successful, it is vital that it is clearly defined as a particular person's responsibility and that sufficient time is set aside to work on the programme. Otherwise it is all too easy for other commitments to take priority.

The RAW programme is offered free to attendees, and any person working with the heritage in any capacity in the West Midlands region is welcome to attend courses. The actual cost of training is about £90 per head. This covers trainer's fees and accommodation, transport and course materials. The courses utilise leading practitioners for each subject area; although the courses are offered free, they are in no way cut-price. The information provided is up-to-date and practical, and trainers have been selected for the quality and breadth of their knowledge, rather than their experience as a trainer. Consequently, some trainers have required more steering than others to help them to provide a high-quality course. This has largely been successful and RAW is now a well-respected course provider within the region, and has also become a course provider for other areas in the UK. More than 200 courses have been given at the time of writing.

IPM courses

One of the most successful courses, in terms of the number of people booking on to the course, has been the pest course. This is run by David Pinniger (latterly assisted by the author) and is split into two separate courses. The first, 'What's Eating Your Collections?', introduces the idea of insect pests and looks at what insects need to thrive, basic identification, pest trapping and monitoring, and has a practical survey exercise in a museum or historic house. The second, 'integrated pest management', includes much more information about identifying insects, including a practical identification exercise, and also looks at control and eradication strategies. There is no requirement for the participants to attend both parts, although this is recommended, and participants are encouraged not to attend part two if they have not attended part one. The key factor is that the course is held in small museums around the West Midlands. This allows participants to train in familiar surroundings and allows them to directly relate the information to their own situation. They can also share their experience with those in similar situations to their own and discuss strategies for managing insects in very practical terms.

Participants receive a manual with a set of IPM guidelines and a copy of the English Heritage/Collections Trust pest poster. All the courses are carefully evaluated to ensure that the courses are fit for purpose and to enable participants to express requests for additional training. The courses are consistently rated as excellent, with the biggest complaint usually being that the room is too cold. Participants comment that the courses are informative and fun, and they go away with a lot of new information and enthusiasm. Because of the way Renaissance is funded, namely directly to the projects, there is some degree of flexibility within the budget to enable us to respond to requests from participants. One such request was to hold a follow-up session to look at new insects that are now being found in museums and to allow the participants to discuss their pest problems with a very experienced practitioner.

Attendance at a training course does not necessarily imply that the knowledge gained will be put to use. Museums applying for Accreditation under the MLA scheme have a requirement to carry out collection care activities, to a degree, or at least to have policies in place to show that they are intending to carry them out. One of the principal aims of the RAW programme was to improve the standard of collection care across the region and this is not solely enabled through training. RAW also offers a free advice and visit service to ensure that information is translated into practice. The advice requested may be as simple as providing the address of a supplier or as complex as advising on how to implement conservation heating. The visits, known as Collection Care Health Checks, may look at issues around a single object or at collection care requirements for a whole museum. Part of the assessment may include looking at the institution's IPM strategy and helping them to make improvements where necessary.

Putting IPM into practice

It may be inferred from the above description that the West Midlands is now a region full of dedicated IPM people, confidently identifying their insect pests and dealing with infestations in a timely manner. Whilst this is undoubtedly the ultimate aim of the RAW programme, it would be over-optimistic to suggest that this was the current situation. To date, 1,192 participants from around 150 institutions have received training in some aspect of collection care. There are 240 museums in the regions, so this figure shows that 62 per cent of museums have received some collection care training. Attendance at RAW courses is not mandatory, and so although we would like to ensure that all museums receive training, we cannot force them to do so. We also cannot guarantee that those who attend completely understand all that they have heard. I visited a museum recently that is extremely good at collection care and there were pest traps set out in their museum at floor level and in corners. I was impressed with their monitoring until I asked them how they recorded their finds. They did not carry out a regular checking regime, nor did they record their finds, nor were traps marked on a plan. I suggested that they might find it helpful to check traps and record finds more regularly, and that they would also find it helpful to note what finds were found on which trap. They could then begin to build up a picture of what insects were being found where. At least this museum had made an attempt at monitoring. Several of the museums I have visited had no monitoring at all, or they may have one or two traps placed on shelves and nothing on the floor. There is still a long way to go.

There have been successes, however …

The Cider Museum, Hereford

The Cider Museum in Hereford had no pest monitoring in place until staff attended a RAW course. Afterwards, they were checking objects during a documentation exercise and found insects on one of the objects. Using their newly-gained pest knowledge, they were able to identify one of their finds as a carpet beetle. This may not seem to be a huge achievement, but for a museum that is staffed by non-museum professionals it was a significant step forward. It was felt to be such an achievement that it was used as a good news story in Herefordshire's Museum Network newsletter (Mayes-Wright 2006).

Shakespeare's Birthplace Trust

More significant were the improvements made at the Shakespeare's Birthplace Trust. The Trust consists of five historic buildings, with period furnishings, a library with a lecture room and a visitor centre. Until Rena Mackenzie attended a RAW course in 2005 there was no insect monitoring in the properties. Rena had had no knowledge of what insect pests were or what to do about them. She attended both parts of the course and used the opportunity to ask a lot of questions. Afterwards, she was able to convince her management team that pest monitoring was required and began a monitoring programme straight away. Rena implemented a perfect example of IPM. She put out numbered traps and marked them on a plan. Regular checking soon revealed that she had a major moth infestation. By carefully recording which traps had the largest catches she was able to track the source of the infestation to some raw wool stored at ceiling height in one of the rooms at the Birthplace Museum. The wool was removed, the surrounding area was sprayed with the insecticide Constrain, to kill any stray moths, and the infestation was eradicated.

Rena then trained the house staff to help her look for pests and they discovered carpet beetles on a pewter plate. A search of the area revealed that the beetles were coming from a stuffed goose. It was felt to be too damaged to save and was removed from the collection. Sadly, Rena is now retiring from the Birthplace Trust, so they need a new champion to build on the foundations she laid.

Developments and Future plans

RAW does not rest on its laurels, however, and the most recent phases of RAW have seen the development of resources to help people with their IPM. 'What's Eating Your Collections?' has been retained as the branding for these resources – the region understands it and it is very self-explanatory. A CD-ROM was produced as the first attempt to generate an interactive resource. The CD has the largest selection of insect pest images available in any IPM resource, insect fact sheets, IPM information, and a decision tree to help the user determine what their insect problem is and what to do about it. A copy of the CD was sent to every museum in the region and has been well received, and is now being used across the UK. Those users who responded to a survey that accompanied the CD described it as easy to use and that it was very easy to find information about specific insects.

The CD is limited in its interactivity and cannot be updated. Consequently, it was decided to further develop the resource and turn it into a website (www.whatseatingyour-collection.com/). This technology has a number of advantages. Information can be linked more easily, allowing the user to work through a series of choices to produce a customised answer to their question. It also allows the resource to be managed and updated so that it stays current and information about new insects can be added. An additional feature is an insect pest recording database. This is currently in a pilot phase, with a small number of museums in the West Midlands participating. The aim of the database is to log quarterly, verified insect pest finds. Part of the reason for doing this is that there is currently no information about insect pest distribution across the UK. The data collected should show where particular insects are found and how, or indeed if, distribution varies across the region and, eventually, the UK. This information should enable collaboration between museums in their efforts to deal with insects pests. It is hoped that it will also help entomologists to map insects across a wide geographical area.

Conclusion

At the time of writing, it is not clear what the future of the Renaissance programme will be or if it will be possible to continue to provide training in the current form. A change of policy by government or withdrawal of funding could undo much of the good work that has been done, but the CD and the website will provide a lasting legacy of help for those implementing IPM in heritage institutions. Our hope is that when someone asks, 'But how do I know I've got pests?' the answer will be *'What's Eating Your Collections?* should be able to help'.

References

Austen, J 1813 *Pride and Prejudice.* London: Thomas Egerton. This is a deliberate mis-quote: my apologies to Miss Austen

English Heritage 2009 'A Helpful Guide to Insects Found in Historic Houses and Museums' (Poster). London: English Heritage and Collections Trust

Mayes-Wright, V 2006 Press release from Hereford Museums Network

Pinniger, D 2001 *Pest Management in Museums, Archives and Historic Houses.* London: Archetype Publications

Porter, G 2002 'Overview of Collections Information and Advice in the Museums Domain'. Unpublished report commissioned by Resource: The Council for Museums, Archives and Libraries, London

What's Eating Your Collection website, www.whatseatingyourcollection.com/ (accessed 4 October 2011)

25

Crazy as a bedbug: the Integrated Pest Management Working Group's development of resources and best practices for the museum community

Rachael Perkins Arenstein*, Neil Duncan†, Lisa Kronthal Elkin‡ and Christopher A. Norris§

*A.M. Art Conservation, LLC, Scarsdale, New York, USA
Rachael@amartconservation.com, www.AMArtConservation.com (author for correspondence)
†American Museum of Natural History, New York, USA, Duncan@amnh.org
‡American Museum of Natural History, New York, USA, lelkin@amnh.org
§Peabody Museum of Natural History, New Haven, USA,
christopher.norris@yale.edu

ABSTRACT

Since its inception in 2002, the Integrated Pest Management Working Group (IPMWG) has focused on providing resources to make it easier for the cultural heritage community to develop and implement integrated pest management (IPM). A history will be presented on the creation of the group from a simple collaboration between two institutions into a large group with approximately 70 members representing over 40 institutions from the USA, Canada and Europe, and an annual meeting that has attracted participants for over eight years. This paper will describe the initiatives that the group pursued and the results, which are available on the website www.MuseumPests.net. The IPMWG can be seen as a model for creating groups to tackle the development of best practices documents in other areas of collections care, so that each institution need not 'reinvent the wheel' as we work to preserve our natural and cultural heritage.

KEYWORDS

Integrated pest management, IPM, MuseumPests.net, working group, best practices

Introduction

As pesticides and fumigants are increasingly limited for institutional use, an integrated pest management (IPM) plan is fast becoming the only viable option for preventing infestations in cultural heritage collections. However, as anyone who has tried to implement or carry out an institutional IPM programme is aware, it is a time-consuming task that is easily shelved in favour of more pressing or easily accomplished projects. As a result, it is not surprising that Heritage Preservation's Heritage Health Index 2005 survey found that approximately 75 per cent of the museum and historical societies in the United States lacked an IPM programme even though 20 per cent considered such a plan an urgent need (Heritage Preservation 2005).

The Integrated Pest Management Working Group (IPMWG), an ad hoc group of professionals from the cultural heritage community and the private sector who have an interest in pest management issues, has been working for almost a decade to produce resources for collection-holding institutions that will enable them to implement IPM. We believe that the group provides a model for other collaborative, cross-discipline groups to develop best practices in the area of preventive care.

History and development of the IPMWG

The roots of the IPMWG lie in a 2001 bi-institutional collaboration between the American Museum of Natural History (AMNH) and the Smithsonian Institution's National Museum of the American Indian (NMAI) to develop a database that could be used to record and map the results of pest trapping. This led to an informal meeting in 2002 with participants from AMNH, NMAI, Texas Tech University, the Canadian Conservation Institute and Zaks Software, who were all working on pest management projects. This initial meeting focused on development of databases with potential for mapping pest activity, identification of essential data fields for databases and the need to survey the community regarding IPM activities and needs. It also led to identification of some fundamental common interests in the area of pest management. It was agreed that contact would be sustained after the meeting and that the possibility of taking these ideas forward through professional societies, such as the Society for the Preservation of Natural History Collections (SPNHC) and the American Institute for Conservation (AIC), would be explored.

One issue that quickly emerged was that disciplinary boundaries that define existing professions may act as a barrier to tackling community-wide problems such as pest management. It was for this reason that, in 2005, the group at AMNH proposed a meeting that would pool resources to tackle IPM issues. Open invitations were sent to the museum and preservation communities via the NHCOLL-L and Cons DistList email lists for a two-day meeting which ultimately gathered 19 people representing 11 institutions from across the USA, Canada and Europe. Since then, the group has held an annual two-day meeting, hosted by the American Museum of Natural History. Work at these meetings has been accomplished by smaller subgroups in five thematic areas:

- Standards and best practices (S&BP): covering IPM policy development, procedures for putting policies into practice, and tools – such as education packages, risk and cost analyses, and scientific studies – for supporting policies and procedures
- Data collection and visualisation: covering monitoring and trapping methodologies, as well as record keeping and reporting, data modeling and database development
- Identification aids: concentrating on imaging and developing resources such as fact sheets, image libraries and identification resources for common museum pests
- Treatments: compiling fact sheets on methodologies, case studies and bibliographies for both chemical and non-chemical treatment options
- Web resources: providing a compilation of useful resources related to IPM and also overseeing the development of the Group's website, www.MuseumPests.net.

Products and resources of the IPMWG

Since 2002 the IPMWG has created several resources to support the development of IPM programmes in the cultural heritage community.

PestList

PestList is an email listserv for IPM-related topics. As originally conceived, the list was a way for participants at the initial 2002 meeting to communicate. Nine years later the list has over 600 subscribers and serves as a forum for discussion of IPM, pest treatment and insect identification, giving people worldwide access to some of the leading experts in the field of IPM implementation, entomology and preservation. Membership is free and open to any interested party.

MuseumPests.net

The MuseumPests website is the main vehicle by which the IPMWG's work is made available to participants and to the wider community. The website was seen initially as a tool to help facilitate products in development by the group. Assets were organised by type (for example, bibliographies, web resources, PowerPoint presentations, templates, procedures, and so forth). Over time, it became clear that this organisational structure was not suited to resource discovery by visitors who were seeking answers to basic questions: 'Why do I need IPM?' 'How do I implement IPM?' 'I found a pest – what is it?' 'I have pests – what do I do now?' In 2009 a professional web designer was hired and the site reorganised with topical presentation of assets organised by the main elements of an IPM programme (preventing access, monitoring, identification and treatment solutions). Website visits have more than doubled each year for the past three years. In the past year the site has received 117,000 visits (9,782 average per month), with 1.24 million hits (103,381 average per month). Peak activity, not surprisingly, is seen in July through November with September 2010 receiving 15,000 visits and 224,000 hits (Fig 1).

Pest monitoring databases

ZPest, a simple, free downloadable programme that organises pest trapping data and presents it in graphical and/or report format was developed for the IPMWG. Other, more ambitious, database programs with mapping components have been developed by individual IPMWG participants based on information they gleaned from their work on the Data Collection subgroup. While not products of the IPMWG, some of these will be available to the museum community.

Best practices documents

Broadly speaking, best practices are commendable actions and philosophies that successfully solve problems, can be replicated, and demonstrate an awareness of professional standards

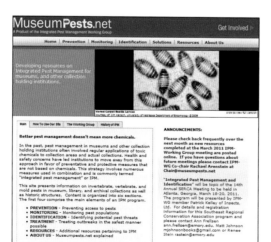

Fig 1
Screen shot of www.
museumpests.net homepage

(Merritt 2008). Codification and development of best practices are generally seen as desirable goals for the museum sector (Macklin 2010). However, there are a number of significant challenges to achieving these goals. To be maximally effective, best practices must be drawn from the widest possible sample of community procedures. The assessment of these procedures needs to be undertaken critically by an adequate cross-section of the professional community. The best practices developed from this process should be made available to the community through publication, either via print or the web. Finally, because best practices are subject to continual refinement and evolution, there must be mechanisms in place for community feedback and regular review.

Traditionally, the role of developing and promulgating best practices has been taken on by professional societies (Macklin 2010) on the basis that these bodies are best placed to access the collective knowledge of their communities and to draw on this knowledge for critical assessment. Best practices that are developed in this way come with a stamp of approval from the society taking the lead, giving them added weight. Under certain circumstances, however, the focus of individual societies may be too narrow to adequately sample the widest possible range of expertise. One such area is that of integrated pest management.

The IPMWG came about because no professional society was able to bring together the varied stakeholders necessary to make an IPM programme a success. The working group encouraged participation from individuals of varied background to ensure that the examples or case studies provided, as well as the tools and resources developed by each subgroup, would explain and demonstrate best practice. Examples of these practices were gathered from the community via member contributions, solicitations of colleagues, and calls for submissions sent out via listservs. These documents were vetted by the subgroups during the annual meeting. Wherever possible, examples were chosen that would have broad applicability, rather than those that were institution- or collection-specific. After obtaining permission from the institutions concerned, these documents were posted on MuseumPests.net along with an abstract written by subgroup members.

Based on the document review, the subgroups generally were able to identify a collection of elements or statements that collectively represented a set of minimum requirements for best practice and from which they generated a template for writing an IPM policy or procedure document. Individual subgroup members were tasked with drafting examples that were then discussed and edited by the wider group. Once these templates were approved by the subgroup they were posted on MuseumPests.net. Examples include the Policy and Procedure Template documents, Treatment Fact Sheets, and Data Fields guidelines. It has been argued by Kronthal Elkin *et al* (2011) that this subset of IPMWG's activities could act as a model for the development of best practices elsewhere in the museum sector. For this to be fully realised, it is necessary to define a set of characteristics for a successful working group.

Development of the working group model

The IPMWG has the potential to act as a broad, community-based model for cross-discipline and cross-institutional collaboration that could benefit other areas of preventive care. Based on the last 10 years, these are the most valuable lessons learnt from IPMWG.

Demand experience

While the invitation to attend IPMWG is an open one, it is always made clear that these are not training workshops and the agenda will not include instruction in basic IPM principles. Early participants were expected to have substantial experience with institutional IPM programmes. While this requirement evolved over subsequent years, it was essential in providing the baseline of expertise necessary to get work off to a quick start

Draw in stakeholders

One of the strengths of the IPMWG is that it draws from the experiences and expertise of a wide range of individuals and institutions; it can truly be seen as a community-led and -supported process. Many different professions, disciplines, and types of institutions are represented by the group's membership. The group has representatives of almost all the stakeholders involved in running an IPM program: these have included conservators, collection managers, curators, and entomologists. In addition, participants with certain forms of specialist expertise were actively recruited when necessary (for example, pest management professionals, facility managers, and so forth). Over the years the scope of institutional types has broadened. Participants for the first few meetings were primarily drawn from natural history museums, but membership soon expanded to include art museums and, by the fourth meeting, there was participation from historical homes, libraries, archives and herbaria. Together they represent a mix of state and federal agencies, universities, stand-alone institutions, and commercial vendors. 'Membership' in the group, defined as someone who has attended a working group meeting or contributed time and resources, is now around 70 individuals representing more than 40 different cultural institutions or companies with continuing international participation. This breadth of expertise is vital for the development of best practices and provides a potential step towards the community buy-in required for the development of professional standards.

Be transparent

The 'About Us' section of the MuseumPests website identifies institutions, companies and organisations involved in the group. Resources submitted by institutions are identified clearly as such, but resources developed for the website by IPMWG members are products of the group and so individuals are not credited.

Develop an open process

With the exception of the first meeting in 2002, and a re-grouping session in 2011 aimed at completing outstanding deliverables, meetings are open to the community. A general invitation is sent out via the listservs and attendees are selected on a first come, first served basis. There is a limit on attendee numbers imposed by the venue and in most years requests to attend have exceeded the number of places available.

Set clear expectations

There is no registration fee for attending the IPMWG meetings but individuals and their institutions cover their own travel and accommodation expenses. The real cost is in 'sweat equity'. Participants in the IPMWG attend with the understanding that they will be leaving with assignments to complete aimed at helping the broader community develop and implement their own IPM programmes. Tasks could take between 24 and 40 hours work over the course of a year. The impetus for most participants, other than altruism, was a desire to learn from what others had already accomplished and hopefully achieve a product that would be impossible alone.

Know your mission

During the first open meeting in 2005 the participants immediately began to develop a framework that would guide the group's work. The first order of business for the IPMWG was to create a mission statement that provided basic direction. Each subgroup was asked to develop concrete short- (one year), medium- (two to three year), and long-term (five to seven year) goals.

Involve the community

At the 2005 meeting a questionnaire was developed and sent out to various preservation and museum email lists to learn more about concerns related to pest monitoring, identification, and data analysis needs. The responses from the approximately 100 respondents guided the work of the IPMWG subgroups, ensuring that products and resources in development would be of interest to the broader cultural heritage community.

Establish a format

By the third meeting in 2006, a format for the two-day gatherings had been established. The whole group would meet for introductions, a review of past activities and the goals for the present meeting. Participants would then break out into subgroup sessions. The groups would reconvene at the end of the day to review accomplishments and assignments as a whole. Because some participants were only able to get travel funding from their home institutions if they 'presented' work, people were encouraged to give presentations during the coffee breaks over the two days. Meetings were held around the same time each year, allowing participants to plan ahead.

Be realistic

After a few meetings a realistic understanding was reached on how much work could be completed during the year. Assignments that people could do on their own – for example, writing case studies or fact sheets, collecting documents, creating bibliographies or lists of web resources – were often successfully completed, but collaborative efforts such as vetting products or developing templates were best done face-to-face at the meetings. After the first three years, established subgroups spent less time in determining what products they wanted to work on and the meeting's schedule was reformatted to provide the groups with time to carry out some of the necessary work there and then.

Create a structure: The IPMWG is an ad hoc organisation with no official status led by a chair or co-chairs and the subgroup chairs. However, the early organisation of the group and continued work running the meetings and website has been facilitated by a small, core group. While these individuals were initially also the subgroup chairs, by 2007 some positions had been turned over to newer volunteers to spread the workload, avoid burnout, and empower the larger group.

Be flexible

IPMWG subgroup chairs were told to use the 2006 questionnaire to guide their work but were given full authority to decide, with their members, their own goals and assignments. It was recognised that IPMWG assignments would fall low on the 'To Do' list when people returned to their institutions and that some deadlines would be missed. When that happened, the subgroup chairs would have to reassess, reassign tasks if necessary, and keep the chair informed on their progress.

Develop collaborative tools: The focus of the IPMWG's work on best practices has been a practical one; to develop tools and resources that can be downloaded and used by any institution.

Show progress

To keep people motivated, email updates are used to publicise progress: for example, when groups complete goals and assignments or when new material is posted to the website. The repeated updating of short- and medium-term goals reminds participants that they are chipping away at long-term objectives and keeps the momentum going. There is no wish to be overly demanding of volunteers and it is recognised that group deadlines will be missed, but numerous missed deadlines without any comment leads participants to feel that the work wasn't really necessary. Follow-up ensures that participants know that their contributions are needed and valued.

Take good minutes

Careful minutes are taken at all meetings. These have proved invaluable in the longer term. On a number of occasions when participants dropped out at the eleventh hour, or when groups transitioned their leadership, having detailed and complete minutes allowed the chair to guide the subgroups back on track and get new leaders up and running.

Be prepared to deal with funding

One of the weaknesses of an ad hoc structure is that at some point any group may need professional services that require payment. The ability to raise, bank, and pay out funds requires, at the minimum, some form of institutional hosting. IMPWG is currently dealing with this challenge, which is likely to fundamentally change the nature and operations of the group (see below).

Future of the IPMWG

As of March 2011, the IPMWG remains an ad hoc group. At an early stage in the development of the IPMWG, the membership debated the issue of whether to seek formal affiliation with a professional society. Ultimately, it was decided that there was no way to do this without potentially limiting either the scope of the resources produced, or the range of membership present in the group. A commitment by AMNH to continue to host an annual meeting has allowed this decision to be deferred by ensuring that consistent logistical arrangements were in place.

The lack of official status became a challenge in 2008 when the group sought to raise funds for developing the MuseumPests.net website. Simply put, dealing with funding required a basic financial infrastructure that the group did not possess. An interim solution was found, but overall there was a feeling that the group might have outgrown its current structure. This led to a formal review of the group's future at the 2011 annual, resulting in a White Paper which will be disseminated to group members in the Spring of 2011. The options under consideration are: disbanding IPMWG; continuing on a limited basis for the purposes of maintaining the MuseumPests site and listserv; or, using these achievements as a foundation for growth.

If the Group chooses to grow and develop, the current operational structure will likely not be sufficient. It would require the development of a formal mission statement and governance structure, and possibly some formal mechanism for institutional hosting. These are substantial issues and are likely to dominate the activities of the group for many months – and years – to come.

Conclusions

The IPMWG provides an example of how a grass-roots, community-led process can cross traditional institutional and disciplinary boundaries, creating a model for developing best practice resources for the cultural heritage community that is broadly applicable for other preventive care issues.

Acknowledgements

The authors would like to thank the IPMWG participants and their sponsoring institutions for their time, expertise and commitment. The sponsors of the MuseumPests.net website are also gratefully acknowledged: American Institute for Conservation – Book & Paper Specialty Group, American Institute for Conservation – Conservators in Private Practice Specialty Group, American Institute for Conservation – Objects Specialty Group, American Institute for Conservation – Paintings Specialty Group, American Institute for Conservation – Wooden Artifacts Specialty Group, The American Museum of Natural History, Artex Fine Art Services, Denver Museum of Nature and Science, Insects Limited, Inc, Museum of Fine Arts, Historic New England, Zak Software Inc.

References

Heritage Preservation 'Heritage Health Index Executive Summary' (2005) www.heritage preservation.org/HHI/execsummary.html (accessed 1 September 2011)

Kronthal Elkin, L, Norris, C and Golpinar, D 2010 'Development of best practices for integrated pest management and a best practices model for the wider museum sector'. *Collection Forum* **24**(1–2), 62–71

Macklin, J 2010 'Best practices'. *SPNHC Newsletter* **24**(1), 6–7

Merritt, E E 2008 *National Standards & Best Practices for U.S. Museums.* Washington DC: The American Association of Museums

26

Skins, shoes and 2,500 saplings: combining integrated pest management and contemporary art installations

Ann French

Whitworth Art Gallery, Manchester, UK, ann.french@manchester.ac.uk
www.whitworth.manchester.ac.uk

ABSTRACT

The Whitworth Art Gallery, University of Manchester, houses designated collections of textiles, wallpapers and fine art, and over the last eight years has been developing an integrated pest management (IPM) programme involving cross-sectional staff training and new working practices. Over the last few years, however, the IPM programme has had to incorporate dealing with hosting performance art events at the Whitworth, and exhibiting contemporary installation works of art. The paper describes the procedures adapted and adopted by the Whitworth to manage the challenges posed.

KEYWORDS

Integrated pest management, IPM, contemporary art installations, performance art

Introduction

The Whitworth Art Gallery (WAG), University of Manchester, houses nationally designated collections of textiles, wallpapers and fine art. Until 2002, only one serious case of pest infestation had been identified that involved the total destruction of a Qashqai nomadic tent, but no formal integrated pest management (IPM) programme existed. The appointment of a full-time textile conservator provided the opportunity to establish a more formal IPM programme for the collections and building. The process began in 2002 with a consultancy survey by David Pinniger of the building and collections that identified priorities on which to base improvements to both building infrastructure and to staff practices.

Since then, the Whitworth has hosted three cross-sectional staff training days (including the Leadership Team) delivered by David Pinniger. These have been followed up by conservation staff who give an introduction to IPM to all new members of staff (including casual Visitor Service Assistants), and new staff accompany the Collection Care and Access Team (CCAT) staff on the trap checking rounds. This programme of awareness raising has been accompanied by introducing new working practices. These include the introduction of a 'food code' that limits the consumption of food by staff and public to specific areas within the building. No staff (or contractors) can consume food of any kind at their desks, but must use the two staff rooms provided. Specific galleries have been identified for corporate hire where the consumption of food and drink is required, such as dinners or gallery private views. Regular liaison with the cleaning contractors enables these areas to be cleaned regularly or specifically after an event. Cleaning methods within the building have also been adjusted and formalised. The CCAT team clean the collection stores, display plinths and cases, cafe staff clean their areas and the cleaning contractors clean the rest of the building on a rota compiled with the help of house services and conservation staff. Dry mopping is being replaced with vacuum cleaning, and more attention is given to nooks, crannies and crevices.

A Whitworth IPM group was re-established in the summer of 2009 after the third all staff training day. An earlier IPM group begun in 2003 had foundered, possibly due to low awareness of IPM and a perception of IPM as too new and seemingly irrelevant. The new IPM group comprises representative staff from teams across the gallery including Collection Care and Access, Curatorial, House Services, Visitor Services, Events, and Learning and Interpretation. The IPM group has enabled communication, discussion and problem solving to find workable practices to improve IPM for the Whitworth. New guidelines, informal and formal, have been produced for events, casual contractors and use of the freezer. Quarantine procedures have been discussed for all teams to encompass new acquisitions, learning materials and lost property. The quirkiest innovation has been the introduction of a compulsory biennial staff de-clutter day to tackle stubborn and long established areas of unsorted clutter that includes filing, out of date marketing material, packaging and similar material, and developing both individual and team responsibility for shared spaces.

Since 2009, the Whitworth has also been working on plans for a Heritage Lottery Funded Capital project that required the compiling of a Conservation Management Plan that included a written formal IPM Policy. This was written in spring 2010 with assistance from David Pinniger and included drawing up plans of the existing building colour-coded as to insect presence risk and compiling recommendations for the architects, MUMA, when planning the extensions and alterations. This policy has been used by the architects and design team when: choosing construction materials; locating ingress points; locating waste and recycling stations; positioning doors and access routes; segregating collection and non-collection areas; and considering areas and routes for food transport and consumption.

New Programming

The outline given above of the Whitworth's IPM methodology is frequently challenged by recent changes in the exhibitions and public programmes staged by the gallery since 2009. The appointment of a new director, Dr Maria Balshaw, in 2006 has encouraged and promoted new approaches to exhibitions and displays. The emphasis is to combine displays of, and access to, our permanent collections of textiles, wallpapers and fine art, with innovative programmes of contemporary art including performance and installation art. Three recent examples will be used to examine the challenges these pose to IPM and how the Whitworth has responded by adapting and evolving IPM practices to cope. While aspects of our programmes are staged elsewhere in other institutions, few seem to combine such programming with a permanent collection so vulnerable to infestation. The policies and systems described by Blyth and Battisson (2008) for the Victoria and Albert Museum are a direct and extremely useful comparison but were developed separately.

The three examples to be cited are *Marina Abramovic Presents …* (3–19 July 2009) for the Manchester International Festival, Olafur Eliasson's *The Forked Forest Path* installed for *The Land Between Us* (25 September 2010–23 January 2011) and *Rotor* by the Siobhan Davies Dance Company (28 January–6 February 2011). The first and most complex of these three was *Marina Abramovic Presents …*, but the experience involved has proved invaluable and precedent-forming.

Marina Abramovic Presents … 2009

For *Marina Abramovic Presents …,* the entire Whitworth Art Gallery was emptied of its regular displays, including 12 dressed cases that form the Textile Gallery. The entire gallery was then given over to 14 performance artists for two weeks. The operational planning for this process was immense and multi-faceted, but only that affecting IPM will be detailed, and only two artists will be highlighted as exemplars for the procedures invented. Each artist invited to perform for *Marina Abramovic Presents …* visited the Whitworth or liaised with the Curator (Modern Art) to establish where and what they were to perform or install, one aspect of which was to discover what materials and staging would be needed, or required to be brought into the building, and whether IPM procedures such as freezing could be applied to any materials belonging to

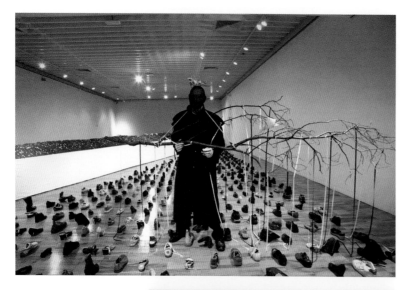

Fig 1 (above)
WARP WRAP, actuation
(performance/installation),
Whitworth Art Gallery,
Manchester, England, 3–19
July 2009, by Alastair
MacLennan.
(Photograph with
permission of the artist and
photographer Marco Anelli
© 2009)

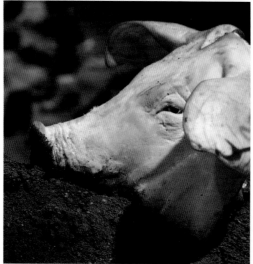

Fig 2 (right)
WARP WRAP, detail of
actuation (performance/
installation), Whitworth
Art Gallery, Manchester,
England, 3–19 July 2009,
by Alastair MacLennan.
(Photograph with
permission of the artist and
photographer Marco Anelli
© 2009)

Fig 3
Rug, Rug, Rug, Rug,
Whitworth Art Gallery,
Manchester, England, 3–19
July 2009, by Jamie
Isenstein.
(Photograph with
permission of the artist and
photographer Joel Chester
Fildes © 2009)

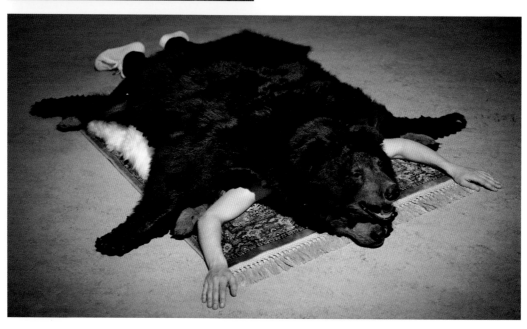

the artists. This process was intricate, diplomatic and time-consuming so as not to intrude on the creative process. In turn, this required creative and rapid problem solving from the Whitworth team including the liaising curator, the registrar, conservators, technicians and volunteers.

Alastair MacLennan's WARP WRAP actuation (performance/installation) required a 12 metre long trestle table covered in a cotton table cloth, piled high with soil on to which were placed two pig heads either end and a number of fresh mackerel (Figs 1 and 2). Surrounding the table 250 pairs of shoes were carefully and strategically positioned, while over a four hour period Alastair MacLennan slowly moved around the trestle table while wearing a coat and stockinet mask over his face and carrying a tree branch. Jamie Isenstein's Rug, Rug, Rug, Rug involved her lying on a carpet under bear, wolf and sheep skins (Fig 3).

All these materials had to be sourced and then a decision made as to whether any further IPM action was required. For Alastair MacLennan, therefore, we discovered that soil is sterilised before purchase and was bought in from a DIY chain, the branch was picked up in our adjoining park and sprayed with Constrain insecticide. The mackerel and pig heads came from our cafe suppliers, but the mackerel was supplied fresh every day and disposed of every day, while the pig heads were returned to the cafe freezer every night.

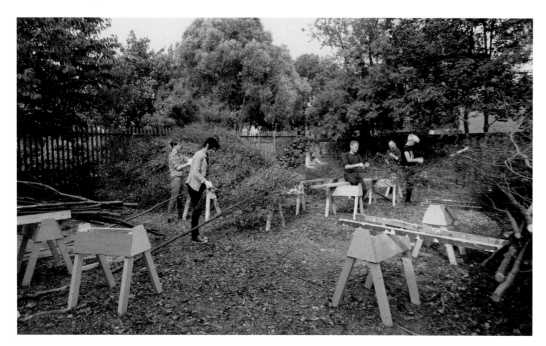

Sourcing the shoes proved more complicated. Appeals to staff and the public produced only 50 pairs, so Oxfam was approached and it made up the surplus. However, given the variety of sources for the 250 pairs of shoes, it was decided that they should be frozen before installation. To make this process as simple as possible, mobile bins were placed behind the information desk and the Visitor Services team were given a supply of zip-lock bags. All staff and public who supplied pairs of shoes were asked to put them into the bags, seal them and then place them into the bins. A team of volunteers dealt with all the shoes from Oxfam in a similar manner. All the shoes were then added to a large consignment of material sent to the Liverpool Conservation Centre for freezing.

Jamie Isenstein's performance props proved more complicated to source as CITES regulations forbade the import of the wolf and bear skins. Alternative UK-sourced skins were found, and again freezing was decided to be necessary. The rug and furs were also all prepared for freezing and added to the Liverpool consignment.

The Forked Forest Path 2010–11

The installation, *The Forked Forest Path*, by Olafur Eliasson involves 2,500 tree saplings wedged between the floor and the ceiling, with paths created to lead the visitor through (Figs 4 and 5). As with *Marina Abramovic Presents* …, the issues were one of sourcing the materials involved, deciding what measures were to be taken and then disposal of the material. An experienced freelance technician, who regularly works with the in-house technician team, was given the task of planning, installing and de-installing the project with a volunteer team under guidance from Conservation and the Curator (Modern Art). Most of the saplings came from Thurstaston Common, a National Trust site on the Wirral. The National Trust rents out some of its land to tenant farmers,

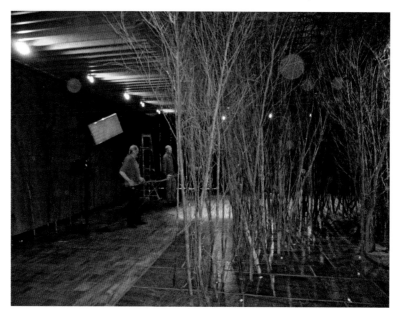

Fig 6
Camouflaged insect blunder
traps inserted at entrance to
Gulbenkian Gallery
containing The Forked
Forest Path.
(Photograph with
permission of the
Whitworth Art Gallery
© 2010)

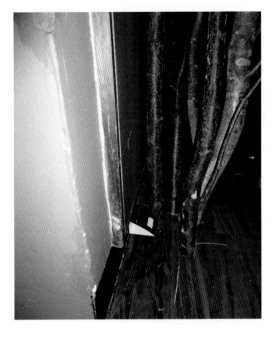

and at Thurstaston they had an area that had been used to grow spruce for use as Christmas trees. About eight years ago, this land was cleared and pioneer trees soon grew into an unmanaged forest. Pioneer trees such as silver birch and sycamore are the first trees that will populate a cleared piece of land. They will self-seed growing in poor soil where other trees, such as English oak or elm cannot. The National Trust wanted to use this land for tenant farmers again and the site had been earmarked for clearing in December 2010. Ordinarily, the trees would have been felled and then burned or chipped and would have seen no further use. They brought the felling date forward to June so we could use them. All of the remaining trees were bought from a coppice cut from a managed forest in North Yorkshire. All 2,500

saplings were prepared for installation on waste ground at the rear of the Whitworth. This involved stripping off all the leaves, spraying the saplings with Constrain insecticide and with a fire retardant Flamebar N5 recommended by the University's fire-officer.

The saplings were all then brought through the gallery and installed in the Whitworth's Gulbenkian Gallery. Insect blunder traps were disguised by painting them black and laid at all the entrances (Fig 6). The reasons for choosing the fire retardant and insecticide were budgetary and for speed and ease of application to meet COSHH regulations. However, disposal of the tree saplings was not fully considered until after installation, which was a mistake, and a lesson learnt (Fig 7). After considerable assistance and calculations by the University's waste management officer, we were allowed to dispose of the saplings in landfill, but the quantities and combination of fire retardant and insecticide came close to declaring the saplings as hazardous waste, which would have incurred considerable cost. In hindsight, the comparative costs of responsible disposal by an approved hazardous waste company versus treatment via heat or low temperature should have been calculated as part of the planning procedures and stands as a future procedural recommendation.

Rotor by the Siobhan Davies Dance Company 2011

Rotor by the Siobhan Davies Dance Company involved a combination of dance performances set against a series of installations involving AV technology, vintage furniture and some textiles. Being the most recent installation, the practices established by *Marina Abramovic Presents …* and *The Forked Forest Path* were applied virtually immediately. Two curators went to view the performances/installations in their London venue, and lists of object props were quickly established and sent directly to Liverpool for freezing. A further challenge was the provision of a Green Room for the dancers. The Worthington Room, adjacent to the main public areas and ordinarily used for learning activities, meetings and small public events, was given over to the artists and dancers. This area is already designated as a high-risk area allowing food consumption, so additional cleaning of the room was added to the cleaning schedules; a fridge and dedicated regularly emptied

Fig 7
Disposal of tree saplings
using an accredited waste
disposal company.
(Photograph with
permission of the
Whitworth Art Gallery
© 2010)

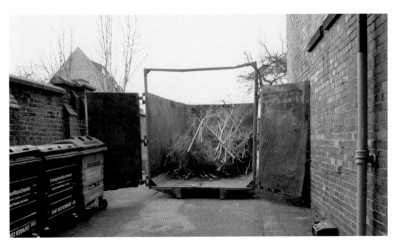

bins provided. Although, certain learning activities had to be moved and/or re-scheduled, the IPM systems remained in place.

New practices

The examples of new programming listed above have provided the Whitworth with much valuable experience to adapt and use for the future. New IPM practices include:

• liaising with artists to establish materials and practice as early as possible

• appreciating that these demands must not alienate creativity

• a rapid and creative response may be required from conservation staff

• dedicated and diplomatic liaison with all concerned works best

• consideration of additional preparatory and storage space may be required

• duration and de-installation practices are as important as installation

• maintain IPM training and awareness across all staff at all times.

The kind of new display and event programming at the Whitworth, as described above, has tested the application of IPM, but not disabled the systems. The still evolving practices have emerged from good staff awareness via IPM training and good communication, and are required at every stage – installation, duration and de-installation.

At the planning and installation phase, the project co-ordinator (in WAG's case normally a curator) needs to establish, carefully and diplomatically, the range and source of any materials required to enter the building. If a new commission is involved, then this can take time and be rather last-minute, therefore a rapid and creative response from conservation is necessary to devise and manage solutions. Contingency must be allowed for in the budget planning to cover these solutions. Consideration of IPM practices has been added to the Whitworth's exhibition planning checklist. Likewise, the needs of artists and performers during the event period must be similarly considered and budgeted, together with any special disposal requirements. Where alternative methods can be applied, for example, heat treatment versus freezing versus insecticide, a cost–benefit analysis is recommended.

To summarise, lessons the Whitworth has learnt are as follows:

• providing all staff and cross-sectional staff with IPM training has raised awareness for all gallery activities

• encouraging planning questions to be asked at an early stage and consulting conservation staff enables solutions to be found for installation, duration and disposal situations.

Conclusion

Contemporary performance and installation art, when combined with vulnerable collections, pose integrated pest management challenges. A vulnerable collection need not be affected, however, as the Whitworth's experience shows, if conservators respond in a creative, positive (and often rapid) manner giving constructive and deliverable solutions. The rule of response is not 'whether' an activity is to proceed but rather 'how' it is to proceed. This is made easier if all staff are aware of IPM issues through training, and in the Whitworth's case, a dedicated cross-team working group. Good communication is essential and all lessons learnt need evaluating to improve future procedures and budgetary considerations.

Acknowledgements

The author would like to thank the following. David Pinniger, whose support, advice and training have been immeasurable; Conservation and Collection Care and Access Team staff at the Whitworth for their help and support in implementing IPM at the Whitworth; Matt Denniss for his supervision of the installation and de-installation of *The Forked Forest Path;* Mary Griffiths, Bryony Bond and Helen Stalker of the Whitworth Curatorial Team for listening and helping to apply IPM solutions; and Liverpool Conservation Centre for use of their freezer.

References

Blyth, V and Battisson, C 2008 'Dangerous liaisons', *in* Saunders, D, Townsend, JH and Woodcock, S (eds) *Conservation and Access, IIC Congress Preprints, London, 15–19 September 2008,* London: IIC, 93–7

27

Where do we go from here? The challenges of implementing an IPM programme at the British Museum

Julie Phippard

British Museum, London, UK, jphippard@thebritishmuseum.ac.uk
www.britishmuseum.org/

ABSTRACT

Although the British Museum was one of the first museums in the UK to accept the importance of an integrated pest management (IPM) approach, the programme never became fully embedded in the institution at all levels and across the whole estate. A number of attempts were made to improve the programme, but it remained somewhat reactive and pest problems continued. Following a recent internal review of IPM, the museum has now assessed its collections and buildings and introduced risk zones to target resources more efficiently. An implementation plan clarifying policies and procedures is in development. Due to the organisation's size and complexity, getting IPM back on track has been challenging. This paper highlights the difficulties in implementing IPM at a large national institution, how these problems are being overcome, and what the future holds for pest prevention at the British Museum.

KEYWORDS

Integrated pest management, IPM, preventive conservation, risk zones

Introduction

The British Museum (BM) cares for a collection of eight million objects spread over an estate that includes an extensive complex of buildings at the main site in Bloomsbury totalling 75,000 m², as well as two off-site storage facilities. There are eight curatorial departments that manage their own collections and storage areas, while responsibility for pest management is split between the Facilities, Buildings Services and Conservation & Scientific Research Departments. Integrated pest management (IPM) was recognised as a practical and useful concept in the museum over a decade ago, but the actual implementation of an IPM programme has proved to be difficult to manage and maintain, leading to a number of pest problems in recent years.

History of pest management at the British Museum

Like many museums, the British Museum historically managed insect pests in its buildings with pesticide sprays. There is little information about what treatments were used on the collections prior to the second half of the 20th century, when records indicate that a hydrogen cyanide gas chamber was used for fumigation at the Bloomsbury site. Due to safety concerns, in the early 1970s a new chamber using ethylene oxide was built at the east London store. During this time, some areas of the estate were also protected by the use of Vapona (dichlorovos-coated plastic strips) until it was banned in 2002. After the use of ethylene oxide was discontinued in the late 1980s, an internal report identified freezing to be the most practical control method (Daniels 1990) and a walk-in freezer was installed in the mid-1990s.

As early as 1991, guidelines were produced in the museum to prevent the biodeterioration of the collections using an integrated approach, including housekeeping, monitoring, environmental control, quarantine and regular staff training (Daniels and Rae 1991). Although a number of these recommendations were outlined again in the museum's 1993 Environmental Policy under the section dealing with insect infestations (Oddy and Bradley 1993), it was not until 2001 that the guidelines were reissued as a more official document on how to implement the policy in general terms (Daniels and Rae 2001).

It is apparent that only some of the measures recommended in the 2001 implementation document were actually put into place, as by 2004 a major clothes moth (*Tineola bissel-liella*) outbreak was reported at the Bloomsbury site, which had started in the upper galleries and quickly spread to other parts of the museum (Korenberg 2004). It was discovered during investigations that the main food source for the moths was wool carpets in the galleries, but moths were also found to be infesting handling collection items. A lack of adequate housekeeping was noted in many areas, with a build-up of organic debris cited as a contributing factor. There was no regular insect monitoring in place and the absence of accurate trapping data made it difficult to determine when the infestation had started.

Poor communication meant that although gallery staff had reported the moths, the information was not passed on to the curatorial or conservation departments for some time. In some parts of the building, the source of the infestation was never determined. A programme of spraying, carpet removal, deep cleaning and freezing of affected objects was undertaken to control the infestation. This was complemented by pest identification training courses for the museum assistants in the curatorial departments, conservators, curators, staff from facilities and gallery wardens, as well as sessions with the cleaners to raise their awareness. However, it was noted at the time that a number of areas were still not treated and that the risk of another serious infestation was high unless an IPM strategy was fully implemented.

Blythe House

Blythe House is a former National Savings Bank headquarters building in west London that has been converted to a collections storage facility, and is shared with the Victoria & Albert Museum and the Science Museum, with each museum occupying its own section of the building. The building has a shared entrance and reception area, and a common mess room. It differs from other parts of the BM's estate in that it has a central building manager, and a separate budget contributed to by all three institutions and administrative responsibility rotates between them. The building has a pest consultant who collates the monitoring results from all areas and produces a report that is discussed at an annual meeting attended by the IPM representatives from each museum. There has been a regular insect monitoring programme at this site since 1997.

In 2003, an IPM strategy for Blythe House was introduced which outlined specific measures to protect collections throughout the building (Pinniger 2003). This was the first document to fully outline tasks required to prevent damage from both insect and rodent pests in an integrated programme of maintenance, monitoring, inspections, housekeeping and other procedures to control the risks associated with incoming materials, food consumption areas, waste disposal, and so on. It was also the first time the need for a pest team was identified with roles and responsibilities outlined. This strategy was fairly effective and the BM's collections at Blythe House remained largely pest-free for a number of years. The success of the IPM programme at this location may be attributed to a number of factors including good communication, regular monitoring and reporting, central co-ordination, a designated budget and the pressure not to pose a risk to the other museums sharing the building.

Orsman Road

The BM's Orsman Road store in east London consists of two buildings separated by a yard. Building one holds five floors of ethnographic collections, containing some objects extremely vulnerable to pests, including those made of fur, feathers, skin, wool and other organic materials. In the late 1990s, the collections were packed in temporary cartons and crates

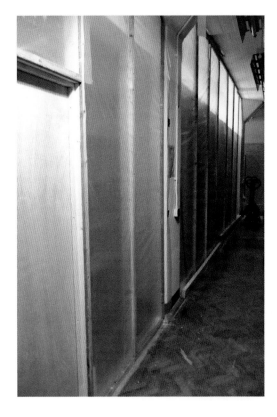

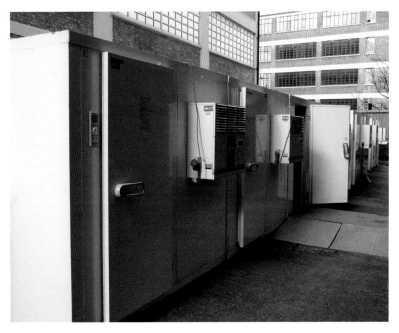

in preparation for a move to a new storage building. This move never happened and the objects remained packed, with some unopened for over a decade. Over time, uncertainty about the future plans for the store, coupled with restructuring and cutbacks in staffing, resulted in poor housekeeping, overcrowded storage areas and inconsistent maintenance. Although there was some pest monitoring, it was not recorded in a co-ordinated manner, so it is difficult to know the exact history of pests in the building.

In the spring of 2008, a webbing clothes moth outbreak was reported in the South American collection on the second floor. This infestation appeared to be concentrated inside boxes containing a costume made of hair, but upon further investigation was found to have spread to other items. In fact, the moths were so established that the larvae had chewed through plastic wrappings and were found in gaps in the metal shelving, which had to be taken apart to be treated.

Unfortunately, this collection was located at one end of a large, open storage room and by the end of the year, it was clear that moths were also infesting the North American collection located at the opposite end of the floor. It was quickly recognised that due to the severity and scale of the infestation, a detailed plan to treat the entire floor with funding as a major BM project was required. A multi-departmental team was drawn together to manage the project and the space was treated in two phases. In phase one, a temporary partition was created, with lobbies added to the lift and the entrance to the room. This divided the floor into two well-sealed spaces that could be accessed separately without risk of cross-contamination (Fig 1). Ten portable freezer units were hired and set up in the yard outside the building to treat the collection on-site. The first half of the floor was emptied and the objects were left in their crates, stacked onto pallets which were wrapped in plastic and then sent to the freezers in the yard for a treatment time of two weeks at −25°C (Fig 2). Extra space had to be prepared in building two to store material after freezing as not all of the pallets could fit into the freezers at once and the treatment had to be done in several batches. While the racking was empty, the space was deep cleaned and then treated with a water-based insecticide called *Tenopa* (Alphacypermethrin 2.97 per cent and Flufenoxuron 2.97 per cent). In phase two, the process was repeated for the second half of the floor. The treatment process cost over one hundred thousand pounds (not including staff time) and took nearly five months, but the actual recovery of the collection, including unpacking, inspection, cleaning and repackaging, will take much longer. There are no formal plans for these recovery tasks to be undertaken in a systematic way as funding for the project only covered the emergency treatment of the collection and storage areas, and staff members who were involved in the project have now gone back to their previous duties.

IPM review

A new pest monitoring programme for the Bloomsbury site was introduced in 2004, and an IPM strategy based on the Blythe House strategy was launched across the museum in 2005, which included the appointment of an IPM co-ordinator. Responsibility for co-ordinating training and monitoring was passed from conservation scientists to organics conservators, who were able to make some progress and practical improvements, but there were still insufficient staff and resources allocated to IPM to implement the strategy. The BM began to recognise the importance of having staff dedicated to preventive conservation, and in 2007 its first preventive conservator (the author) was appointed to co-ordinate the IPM programme, with a second preventive conservator appointed in 2009.

In response to the Orsman Road infestation, and a number of ongoing moth problems across the Bloomsbury site and more recently at Blythe House, a review of the IPM programme was initiated with the following remit:

1. to review all of the documentation pertaining to pest management to determine if the museum's IPM strategy was sound

2. to establish what current procedures and facilities were in place, in order to determine why the IPM programme was not successful

3. to make recommendations for improvements and create a detailed implementation plan to get IPM back on track, and to raise awareness throughout the museum at all levels.

In stage one of the review, the review team determined that the various documents pertaining to IPM contained all the main components for an effective IPM programme but that they lacked the level of detail necessary to actually implement them. There was some confusion over the purpose of each document and it was unknown how widely some had been circulated. The review team recommended that the museum produce and adopt three distinct documents to support the IPM programme:

1. an IPM policy, to state the museum's commitment to IPM, forming the guiding principles, aims or rules governing IPM

2. an IPM strategy, to outline the general approach to IPM across all sites, in order to fulfil the aims of the policy

3. an IPM plan or manual, to contain specific information and procedures on how to implement IPM museum-wide and at both outstations. It should contain detailed sections outlining roles and responsibilities, step-by-step procedures, pest identification aids and other practical information to serve as a reference and training tool for all employees.

The review team proposed a new IPM policy for the BM, which was adopted soon after (British Museum 2010). This was seen as an essential step as this policy would underpin any other work undertaken to improve the IPM programme.

During stage two of the review, the team liaised with various departments to find out what IPM procedures and facilities were in place, and to understand what factors were preventing IPM from being successful. The team discovered that most curatorial departments did not have adequate resources to perform IPM-related tasks. For example, few departments had any designated quarantine or inspection space in which to check incoming objects before putting them into storage. Most teams of museum assistants had very little time to undertake regular, routine deep cleaning of the storage areas, or else the stores were so overcrowded as to make deep cleaning nearly impossible. Although insect identification courses had been running for several years, many staff struggled to find the time to do the quarterly trap checking, or this was not seen as a priority. Awareness of IPM issues in non-curatorial departments appeared to be nearly non-existent.

An inspection of the insect trouble spots by the museum's consultant entomologist in 2009 highlighted a number of issues including poor communication between departments, complicated budget arrangements for IPM tasks, lack of ownership of some spaces, and uneven response to pest problems. In essence, IPM was still too reactive and not adequately co-ordinated with other museum programmes such as maintenance, refurbishments, housekeeping, exhibitions and events. This related directly back to the review team's findings that recommendations and lessons learned from previous infestations had been implemented in an inconsistent way, and without the necessary changes at more strategic levels. One of the main problems was thought to be the absence of an IPM champion at high level to ensure that pest prevention measures were embedded in the museum's practices and a dedicated manager at operational level with enough authority to co-ordinate the implementation of agreed IPM policies and procedures across the museum.

Current status

In light of recent activity, there is a new level of pest awareness at the Directorate level. Pest infestations are now recognised in the business continuity plan, thus ensuring that an organised response and funding can be made available in the event of an emergency infestation in the collection. However, the focus of the IPM programme is to prevent problems and efforts are being concentrated in this area.

Stage three of the IPM review is nearing completion and a detailed implementation plan is in development. In order to effectively target limited resources to the most vulnerable areas and truly understand the implications of museum-wide policies and procedures, the estate is being divided into IPM risk zones. This is a management system that has already been used successfully at other museums in the UK. Some work was done by previous IPM co-ordinators but the current system is far more comprehensive, and includes every space in the estate and not just collection areas. There are separate risk categories for insects and rodents, and a simple traffic light method is used to

Table 1 Insect (IN) and rodent (ROD) risk zone categories at the British Museum

Risk zones – insect

Colour code	Risk code	Risk zone	Description of zone
	A-IN	Very high	Very vulnerable material in storage (includes handling collections). Very vulnerable material on open display (temporary or permanent). Very vulnerable material on display in areas that have a history of infestation. Any actively infested area.
	B-IN	High	Very vulnerable materials on display in all other areas.
	C-IN	Medium	Very vulnerable material in transit or on temporary display (less than 6 months). Vulnerable material on display or in store.
	D-IN	Low	Galleries and stores with no vulnerable material and no other risks. All other areas.
	E-IN	Caution	Galleries/stores with no vulnerable material but have other risk factors (e.g. floor grates or other debris-collecting voids, wool carpet) or a history of infestations. Non-collection areas that pose a risk to adjacent collection areas (e.g. areas with wool carpets, storage rooms with vulnerable material, some departmental kitchens).

Risk zones – rodent

Colour code	Risk code	Risk zone	Description of zone
	A-ROD	Very high	Areas where food is regularly stored, prepared, served, consumed or disposed of. Areas with current or recurring rodent problems.
	B-ROD	High	Areas where there is public access.
	C-ROD	Medium	Occupied areas (offices, workshops), spaces with water source (toilets), circulation spaces (loading bays, lifts, lobbies, basement corridors) unless well proofed.
	D-ROD	Low	Museum object and other storage areas (as long as they are adequately proofed). All other areas (including plant rooms, internal corridors).

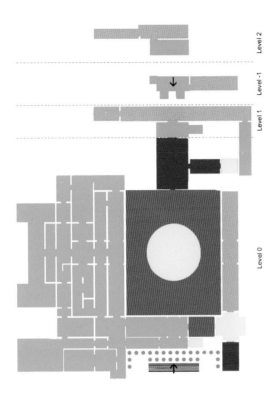

Level 2

Level -1

Level 1

Level 0

Level 1

also helping to identify where populations are established in voids and other inaccessible areas creating a reservoir of pests that pose a continuing risk to collections. Some of these areas were not previously monitored and therefore the risk was unrecognised.

Fig 3
A floor plan of the lower British Museum galleries, showing the insect risk zones.
(© Trustees of the British Museum)

Where do we go from here?

Developing the policies and procedures for each risk zoned space is a complicated undertaking, which must be done with input from relevant stakeholders within the museum in order to make them practical and sustainable. A number of programmes are in development to re-launch IPM, from training courses for the cleaners to presentations to raise awareness across the museum, to induction materials to ensure new staff are aware of their responsibilities relating to pest reduction. IPM will not be successful in the long-term without the co-operation of all staff.

colour-code different zones from high to low risk. The system is being used to recognise the different factors that can affect the risk rating of an area, including the vulnerability of the collection, the potential of the environment to harbour pests, the history of pests in the area and the use of the space (Table 1).

Mapping the risk zones on floor plans shows the relative risks of adjacent areas and allows for strategies to be developed to deal with access, circulation routes and special events (Fig 3). Each risk zone has corresponding standard operating procedures so that appropriate maintenance, housekeeping and monitoring programmes can be tailored to each zone. With over 200 storage areas, approximately 80 galleries and probably thousands of other occupied and circulation/service spaces across the estate, assessing risk zones is no simple task! The zone ratings should also be updated regularly to reflect changing use of spaces, redevelopments and building works.

Improvements in the monitoring programme to incorporate risk zone information, and the availability of better and cheaper monitoring tools such as moth pheromone lures, have resulted in a much better picture of insect population levels and locations. This has shown that sometimes problems develop in non-collection areas adjacent to collection areas. It is

IPM is now built into department plans and thus from the 2011–12 planning year there will be quarterly performance targets reported to the BM Trustees. IPM is also being better tied in with other museum-wide programmes such as waste reduction, vertebrate management, stores improvements and gallery refurbishments. The BM is in the process of building a new wing at the Bloomsbury site and a logistics hub complete with a new pest treatment facility is planned in the main receiving area. This will provide a central inspection, quarantine and handling space for all departments, ensuring that the risk of contaminated materials entering the collections is greatly reduced. Material from Orsman Road will be relocated to two new storage levels in the basements, and careful planning will be needed to ensure that no pest problems are imported during the move.

One of the key areas currently under debate is how the IPM programme will be managed and funded in the future. Past problems have highlighted the need for strong leadership both at managerial and operational levels, and IPM is so important that it cannot simply be tacked on to the responsibilities of an existing staff member if enough time and attention is to be paid to the programme. Discussions are in progress to determine if a post such as IPM Manager can be created, but in these times of austerity this is a complicated decision. In any case, the IPM team will be reconstituted with clarified roles and responsibilities, and IPM representatives from non-collection departments recruited to facilitate more coherent pest prevention across the whole museum.

Conclusion

The British Museum is a complex institution spread across a large estate, which has resulted in difficulties with the implementation of an IPM programme in the past. A lack of consistency between sites, poor communication between departments and uncertain leadership of the programme led to a number of pest problems that were managed in a reactive way. A new understanding of the risks that pests pose to the collection, combined with better tools to target resources, have led to the development of more detailed policies and procedures to better protect the collections. Renewed support for IPM at the highest levels of the museum will ensure that pest prevention is more integrated into all the museum's activities and make a stronger and more sustainable IPM programme for years to come.

Acknowledgements

The author would like to thank all of her colleagues in the Conservation and Scientific Research Department, and across the museum, who have offered input and helped to improve the IPM programme. In addition, a huge thank you to the many colleagues at institution throughout the UK who have freely given advice and shared information about IPM at their own sites. Finally, a very special thank you to David Pinniger, whose unfailing support and words of encouragement have made both this paper and the Pest Odyssey conference possible.

References

British Museum 2010 'Integrated pest management policy for the British Museum', British Museum policy document

Daniels, V D 1990 'Methods for the eradication of insect pests in museum objects'. Unpublished British Museum internal report 1990/8, Department of Conservation and Scientific Research

Daniels, V D and Rae, A 1991 'Guidelines for the prevention and control of biodeterioration'. Unpublished British Museum internal report 1991/31, Department of Conservation and Scientific Research

Daniels, V D and Rae, A 2001 'Implementation of the policy on the prevention and control of biodeterioration'. Unpublished British Museum internal report, Department of Conservation and Scientific Research

Korenberg, C 2004 'Moth outbreak in the British Museum'. Unpublished British Museum internal report 2004/10, Department of Conservation and Scientific Research

Korenberg, C, Delaunay, H and Pinniger, D 2004, revised 2006. 'IPM strategy for the British Museum'. Unpublished British Museum internal report, Department of Conservation and Scientific Research

Oddy, W A and Bradley S 1993 'Environmental policy for the preservation of the collections in the British Museum'. Unpublished British Museum policy document, Department of Conservation and Scientific Research

Pinniger, D 2003 'IPM strategy for Blythe House'. Unpublished British Museum internal document, Department of Conservation and Scientific Research

Abstracts of Poster Presentations

Insect pest management initiatives past and present at the Horniman Museum

*Louise Bacon, Julia Gresson and Charlotte Ridley

*Horniman Museum & Gardens, 100 London Road, Forest Hill, London SE23 3PQ, UK,
lbacon@horniman.ac.uk, www.hornimanmuseum.ac.uk (author for correspondence)

The Horniman Museum and Gardens in South East London were opened to the public in 1901. As a gift from tea magnate Frederick Horniman to the people of London, the building, collections and grounds were to be used for recreation, education and enjoyment. The extensive gardens surrounding the museum have made control of pests in the Horniman collections of natural history, anthropology and musical instruments a major issue for over 100 years.

Since the appointment, in 1967, of the first professionally qualified conservator, it has been important to identify past pest management initiatives and to introduce new protocols for the future. This is essential for documentation purposes, and to ensure that health and safety concerns are addressed, and legislation adhered to.

Empty net bags found in storage boxes and hanging in showcases are evidence that moth balls and para-dichlorobenzene

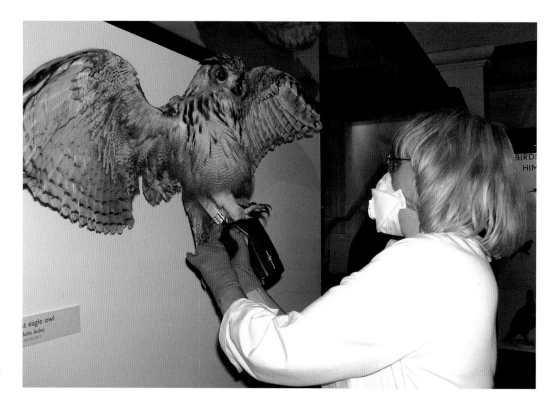

Fig 1
Assessing taxidermy with an XRF hand-held analyser. (© Horniman Museum)

(1.4-dichlorobenzene) had been used. Yellowing and brittle pieces of fabric indicated the remains of dichlorvos (2,2-dichlorovinyl dimethyl phosphate, DDVP). Later, the use of fumigants such as ethylene oxide and methyl bromide were better documented. Since 2005, current practices include targeted use of low hazard residual insecticide sprays, low temperature pest control and nitrogen gas.

Recent investigations have also included a conservation assessment of taxidermy specimens, many dating from the early Horniman period, utilising a portable X-ray fluorescence (XRF) hand-held analyser (Fig 1). This has enabled the detection of a range of chemicals used as preservatives, in particular arsenic and mercury applied to skins prior to mounting – an unpleasant but early example of pest management techniques.

As a result of this work, for many years integrated pest management (IPM) protocols have been in place for processing incoming and outgoing objects, undertaking regular inspections of the collections, deployment of sticky traps and in particular, training of all staff in pest awareness. Towards this end, every year a pest trapping and identification day is held to induct all new staff and volunteers into the IPM policy and practices of the museum (Fig 2).

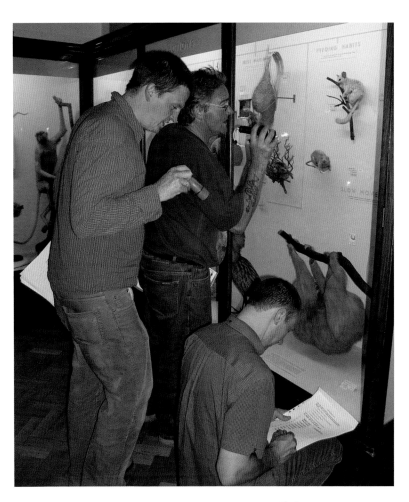

Fig 2
Museum staff searching for insects during a Pest Day training exercise.
(© Horniman Museum)

Pest identification: a tool for training and everyday use

Anna Barnes* and Charlotte Lester

*Conservator Museum, Galleries & Heritage, Environment & Development Services, Rotherham MBC, anna-barnes@rotherham.gov.uk (author for correspondence)

Introduction

Successful integrated pest management (IPM) requires a collaborative approach by museum staff (Lauder and Pinniger 2006). Many people in museums, who are not usually based within collections/conservation departments, are keen to become involved in collections care. IPM is the perfect area in which to develop this opportunity, improving the care of the collections, whilst bringing together sections of the museum that do not always work together.

The challenge

Staff and volunteers that wish to become involved in IPM may only have limited access to relevant information. For example, café staff may never enter back-of-house areas where this information is kept; visitor services assistants may not know which sources are best to help them identify a potential pest; or volunteers may not have the confidence to ask questions about IPM. Enthusiasm needs to be supported in order to develop confidence and ensure the success of a collaborative system.

The concept

In order to encourage task ownership and build confidence, a portable identification tool has been developed. This is a quick reference guide for people who are unfamiliar with IPM, and which can also be used as a refresher training tool. Initial tests with visitor services, café and cleaning staff at Clifton Park Museum (CPM) in Rotherham, proved that staff were enthusiastic about being involved in collections care, an area they had previously seen as 'specialist' and not for them. A series of pest identification cards were produced in a key ring format. Each card is no larger than a credit card and bears an image of a common pest, a short description about the type of materials it eats and the key signs of infestation if the pest itself is not obvious.

Implementation

This key ring, in conjunction with the English Heritage poster (English Heritage 2009) was used to train staff at CPM and develop their IPM knowledge. It raised awareness, not only of the risk to collections, but also of the importance of an integrated approach. The training was especially well received by café staff, who said that at times they felt isolated from the museum (Fig 1). Rotherham Archives and Local Studies are now in the process of adopting this portable identification tool to support in-house refresher training.

Further development

This tool helps to make IPM accessible to all, allowing staff to engage with and support on-going collections care. The simplicity and approachability of the tool makes it very versatile. It could, for example, be applied in international training programmes by creating cards tailored to the specific needs of institutions in a particular region. The potential expansion of this concept to include other collections management issues is being explored and discussed with colleagues.

References

English Heritage 2009 Insect Pests found in Historic Houses and Museums (poster). English Heritage

Lauder, D and Pinniger, D 2006 'English Heritage Guideline for Insect Pest Management in English Heritage Historic Properties. English Heritage'. www.english-heritage.org.uk/content/imported-docs/f-j/guideline_insect_pest_management_at_eh_historic_props.pdf (accessed 15 February 2011)

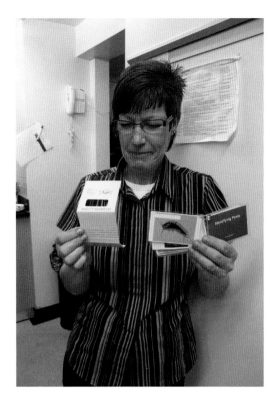

Fig 1
Checking the identity of an insect on a trap by comparing it with the image on the keyring._

Using Killgerm AF *Tineola* pheromone lure boards to target cleaning and intervention at the Imperial War Museum

Philip Baxter

Care Project Manager, Department of Collections Management, Imperial War Museum,
Lambeth Road, London, SE1 6HZ, pbaxter@iwm.org.uk

Background

The Imperial War Museum (IWM) started using Killgerm AF *Tineola* pheromone lures in 2008. Since March 2009 these have been the only pheromone lures used at the IWM as part of the integrated pest management (IPM) programme.

Methodology

Killgerm AF pheromone boards were cut in half and placed on existing blunder traps, these

Fig 1
Secret War Gallery moth catch spread.

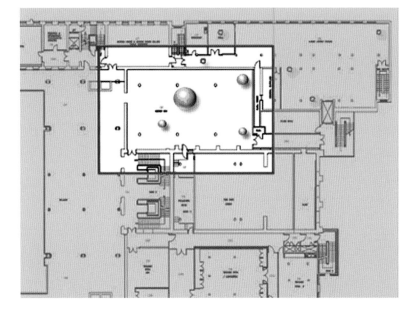

were further supplemented by an additional 66 AF traps with full boards, located in permanent galleries on a broad grid pattern.

Results

After several months of data collection it was becoming clear that certain traps were catching significantly more moths than others within the same galleries (Fig 1).

Working from an assumed principle that moths attracted by the pheromone would be drawn to the nearest lure, the spaces adjacent to the most popular traps were investigated. In some cases obvious harbourage for moths were identified, for example, theatrical blackouts forming part of the gallery design. These were scheduled for removal.

When no obvious harbourage could be identified, further investigation revealed large amounts of organic debris (mostly fluff and dust) behind normally inaccessible panels and in voids. Analysis of samples under a microscope revealed the presence of moth larvae head capsules and frass.

Evidence of moth populations living in organic debris was discovered in several locations in IWM London and also at other IWM sites. This seemed to demonstrate that a sufficient build-up of organic debris could support a viable moth population. Cleaning out of accessible voids and cavities was undertaken.

Following the cleaning of these cavities in winter 2009/spring 2010, there was a

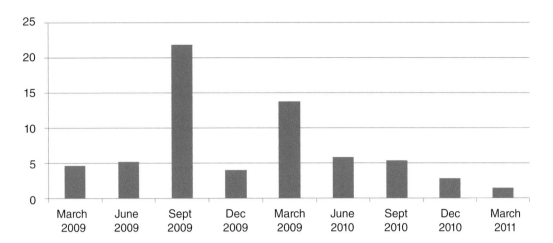

Fig 2
Secret War Gallery traps
average catch.

reduction of activity compared with the previous year, particularly during the summer months (Fig 2). This is most likely attributable to a combination of the removal of eggs during the clean and the reduction of viable harbourage for the next generation to emerge from. Based upon this outcome a further round of deep cleaning is being planned for delivery during 2011. It is hoped that this will further help control the moth population.

Killgerm AF pheromone boards do not provide a magic bullet, however, when used as part of an IPM programme they can be a relatively low cost means of pinpointing where potential sources of infestation may be found. This allows for scarce resources to be more effectively deployed. It also aids our goal for long-term preservation of the collections in our care and underlines the importance of sustaining simple measures such as housekeeping.

Mind the gap! One year on in the trial of a floorboard gap sealer

Artemi Chaviara

MA Preventive and Object Conservator, chaviaraa@gmail.com, Athens, Greece

'Stop-gap', a commercial product, has been used to protect the Melville State Bed, a 17th-century bed and furnishings, from insect damage (Chaviara *et al* 2009). The bed is displayed in Gallery 54A in an enclosed glazed room standing on a raised floor of laid wooden timbers. The bedroom is located within the British Galleries of the V&A Museum (Fig 1).

This four-poster bed is constructed of oak and pine and furnished with embroidered hangings of crimson velvet and ivory silk. In 2008 carpet beetle, *Attagenus smirnovi*, larvae were observed in the room. It was thought that they were living in a dead space under the raised floor as some larvae were caught in the blunder traps and live larvae were found grazing on animal glue on the lower valance of the bed.

Wooden floorboards, original or reproduction, are often used in country houses, historical properties and museum galleries. Gaps between floorboards can accumulate dust, clothing fibres, human hair and skin, providing an ideal environment for insect activity. Some methods used for sealing the gaps in the past have been unsuitable because of aesthetic or practical reasons.

Research for a suitable method of sealing floorboard gaps was undertaken in 2009. 'Stop-gap', a thin, opaque and flexible strip, was considered a possible solution (Fig 2). A sample of the product was sent for Oddy testing to verify its suitability for use in historical settings. Attenuated Total Reflection Fourier Transform Infrared Spectroscopy

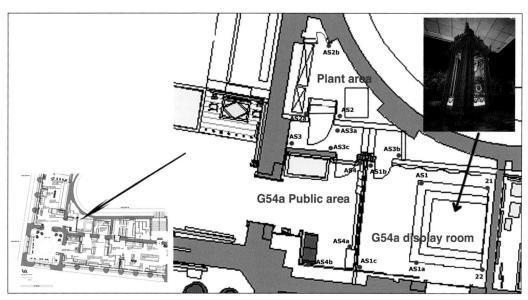

Fig 1
British Galleries map with detail of Gallery 54a and the Melville bed. Red spots indicate blunder trap locations.

analysis confirmed the material's composition as polypropylene; results suggested the sealer had potential for permanent use.

A sample of 'Stop-gap' was applied to gaps in the floorboards in the glazed room in April 2009 to test its performance, quality and suitability for minimising insect activity as well as reducing the accumulation of dust. After a year, quarterly trap checks provided encouraging results for the use of 'Stop-gap'. Data showed a significant reduction in insect activity within the room display which was in direct contrast to an increase in activity in an adjacent plant room area.

As this product has proved effective it may be considered for further use in display areas. This project is still in progress and results will be published when more insect activity data has been collected and conclusions reached on the product's performance in the museums display areas.

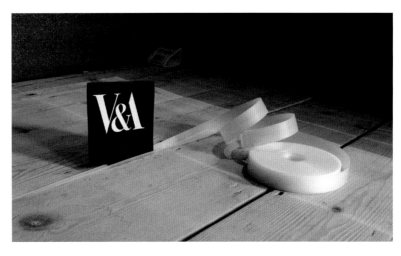

Fig 2
'Stop-gap' is pushed into the gap, using a plastic slide.

Acknowledgements

The author would like to acknowledge the help of the following: Val Blyth, Senior Preventive Conservator, V&A Museum, UK; Suzanne Smith, Display and Collections Management Curator, V&A Museum, UK; David Pinniger, Insect Pest Consultant to the V&A Museum, UK; Boris Pretzel, Materials Scientist, V&A Museum, UK; Julie Phippard, Preventive Conservator, British Museum, UK; and William Lindsay, Post-Doctoral Research Associate, the National Archives, UK.

Reference

Chaviara, A, Blyth, V, Smith, S 2009 'A trial use of a commercial material for sealing the floor gaps in a display case'. *ICON News,* **26**, 33

Desktop pests: treating an early computer to prevent risk to the Science Museum's collection

*Fran Coles and Natsumi Henzan†

**Conservator, Science Museum, Blythe House, London, W14 0QX, England, UK,
fran.coles@sciencemuseum.org.uk (author for correspondence)
†University College London, London, England, UK*

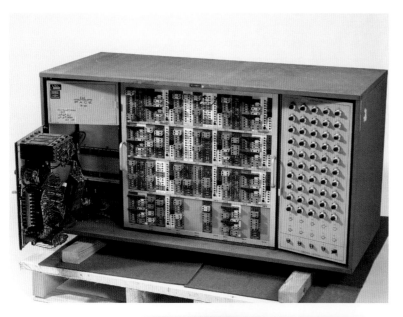

Fig 1
2008–4 PACE computer before treatment.

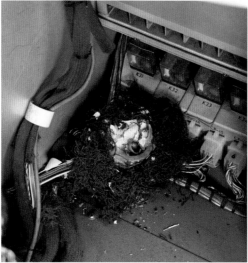

Fig 2
Close up of the rodent nest.

Background

An early example of a desktop computer, the EAI Pace TR-48 Analogue Computer, was acquired by the museum in 2008 (Fig 1). On arrival at the Science Museum storage facility at Blythe House, it was noticed during its pest inspection, that a small rodent nest had been made within the internal workings of the computer (Fig 2). The object was subsequently quarantined, because of the risk of insect infestations, while a treatment proposal was drawn up.

The computer is constructed of enamelled metal, metal components including wiring, plastic coatings (possibly PVC), hard plastic components and woven fabric sheaths. There are a number of compartments and voids within the body of the computer. The dimensions are 1,230mm × 610mm × 510mm.

Treatment methods considered

- Freezing. This is a relatively quick method but because of its size the object would require a walk-in freezer. Things to consider include the protection of delicate components from condensation and embrittlement or shrinkage of the plastics during the process (Shashoua 2008).

- Nitrogen (anoxia). This requires more time, specialist knowledge and equipment. Therefore it is potentially more expensive. There is minimal risk to the object from this method (Maekawa 2003).

- Carbon dioxide. This can be cheaper than nitrogen treatment and has been used successfully in the past for this type of material.
- Fogging with insecticide. This was not considered because of the risk posed to the metal components within the computer.
- Monitoring and deep cleaning. This would be cheap but is not a quick answer. The method would not be suitable for an object that has already shown signs of active infestation.

Outcome

A condition risk assessment, looking at the object construction and materials, showed that this object was at low risk from insect pest attack. Preliminary monitoring of the object had not discovered any pest activity. Combined with budget constraints and a workable time frame it was decided to go for the option of monitoring and thorough cleaning.

The nest and debris were fully documented before cleaning took place. The computer was dry cleaned using microfibre cloths, brushes and a vacuum. Wet cleaning of components to remove residues was then completed. This was done for health and safety reasons as well as for the preservation of the object. Insect pest traps were then placed in and around the computer before it was wrapped in polythene for monitoring.

Conclusion

This treatment raised a number of questions and allowed us to research into the use of traditional pest eradication methods for modern materials. However, it was these modern materials themselves that ultimately allowed us to go for the simplest option – that of cleaning and monitoring. We were aided in this decision by our time frame and budget constraints. However, this will not always be the case and I feel confident that we are now equipped with the knowledge to allow us to make further treatment decisions in the future.

References

Maekawa, S and Elert, K 2003 *The Use of Oxygen-free Environments in the Control of Museum Insect Pests,* Los Angeles: The Getty Conservation Institute

Shashoua, Y 2008 'Inhibition of degradation by low temperature storage', *Conservation of Plastics: Materials Science, Degradation and Preservation.* Oxford: Butterworth-Heinemann, 202–07

Museum pest and IPM reference database

Amy Crossman* and David Pinniger†

*Care of Collections Officer, National Army Museum, Chelsea, London SW3 4HT,
acrossman@nam.ac.uk
†DBP Entomology, Cookham, Berks SL6 9DE, pinniger@globalnet.co.uk

There are a very large number of references relevant to integrated pest management (IPM) which have been quoted in books and papers over the last 20 years. They are in a wide range of publications, some of which may not be easily discovered or tracked down. The objective of this project is to compile a comprehensive and readily accessible reference database system which would include as many relevant sources as possible. An additional aim is to eventually include an overview of the content of the references and an assessment of their current relevance and applicability.

The database is on an Excel spreadsheet with each reference arranged according to the Harvard system (Fig 1). There are additional categories of subject, type of publication and language. Initially, the references were abstracted from some of the key recent IPM publications, such as Florian 1997 and Pinniger 2001. They have since been supplemented with references from many additional sources. The current file is small enough to send by email, but as it grows it will need to be based on a website so that it can be updated and accessed. It is hoped that other IPM workers will share this database and add references from their libraries and collections so that it can be a truly international resource.

Fig 1
Screenshot of database.

Author	Co-authors	Date	Category	Title	Typ	Specific Reference
Abdel-Rahman H A	Soliman Z A and Ali M F	1981	Biology	Biological study on the black carpet beetle, Attagenus scalaris Pic (Coleoptera: Dermestidae).	Paper	The Entomological Society of Egypt 63 1980. pp. 231- 241.
Ackery P	Testa J M, Ready P D, Doyle A M and Pinniger D B	2003	Treatment	Effects of high temperature pest eradication on DNA in entomological collections.	Paper	Studies in Conservation 49, pp. 35- 40.
Ackery P	Pinniger D B, Doyle A and Roux K	2005	Treatment	Heat treatment of entomological drawers using the Thermo Lignum heat process.	Paper	Collection Forum 19, pp. 15- 22.
Ackery P R	Chambers J and Pinniger D B	1999	Monitoring and trapping	Enhanced pest capture rates using pheromone-baited sticky traps in museums stores.	Paper	Studies in Conservation 44, pp. 67- 71.
Ackery P R	Doyle A and Pinniger D B	2002	Treatment	Safe high temperature pest eradication · Is the answer in the bag?	Paper	Biology Curator 22, pp. 13- 14.
Adams P H	Warren R and Horler D F		Monitoring and trapping	Attractancy of long chain fatty acids to Anthrenus verbasci.		Chemical and Biochemical Studies, pp. 207.
Adams R G		1978	Biology	The first British infestation of Reesa vespulae.	Paper	Entomologists Gazette 29, pp. 73- 75.
Adriaens A		2005	Conservation	Non-destructive analysis and testing of museum objects: An overview of 5 years of research.	Paper	Spectrochimica Acta Part B: Atomic Spectroscopy 60, pp. 1503- 1516.
Akerlund M		1991	General	Angra-finns dom...?	Book	Svenska museiforeningen. Uppsala.
Akerlund M	Flato S and Hellekant A	1998	General	Fran silverfisk till halsorisk.	Book	Stockholm, Swden: LTs.

References

Florian, M L E 1997 *Heritage Eaters – Insects and Fungi in Heritage Collections.* London: James and James,164

Pinniger, D B 2001 *Pest Management in Museums, Archives and Historic Houses.* London: Archetype Publications, 115

Ask the Conservator – A pest quest

Rebecca Duncombe*, Suzanna Neil and Victoria Richards

Conservators, Historic Royal Palaces, 0203 166 6436, *rebecca.duncombe@hrp.org.uk

The conservators at Historic Royal Palaces (HRP) design and deliver a wide range of public engagement events at peak times throughout the year to communicate conservation work to visitors. In recent years, this developed into 'Ask the Conservator,' a programme of events spread across the five palaces within HRP's care. One of the most popular 'Ask' events, which spans all the disciplines of conservation, is 'Bugs'. This is a particularly popular subject with family groups (Fig 1).

To build the capacity of this popular event, HRP produced a family-friendly bugs-themed portable exhibit. When manned, it encourages visitors to talk to conservators, asking questions about pest management, either in the palace or in their own homes. When the stand is unmanned, it is an interactive exhibit informing the visitor of the important unseen work carried out by the conservation team.

The portable exhibit was also designed to help tackle language issues with our overseas visitors and younger visitors, through oversized props and an interactive microscope. This allows everyone to take away some basic information of what the conservation team do to deal with the increasing threat of pests to our collections and historic interiors.

The material content of the exhibit is very portable and can be transported and set up by one person anywhere. A visual back-drop of a pest's-eye view of the world sets the scene, and the content covers the damage pests cause to different objects, how we identify them and how we get rid of them. Conservators explain to visitors that they can find the pests in their own homes, and describe cleaning methods to help eradicate them.

This is all communicated using a combination of big, bright, highly visible props, such as a large blunder trap and chemical bottle, and direct interaction with the visitors (Fig 2).

The highlight of the show is a video microscope allowing the visitor to use the microscope to view live and dead examples of the common insects that we get within the palaces. They can also watch the samples via the television

Fig 1
Ready to delight the customer at an 'Ask the Conservator' bug event.

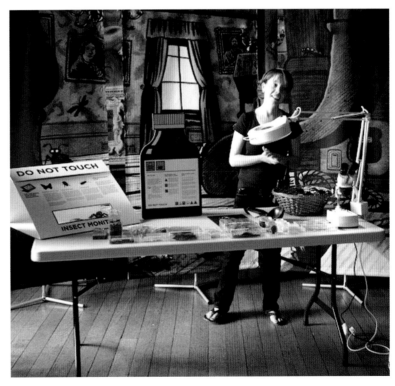

Fig 2
Oversized pest trap for the
public to view.

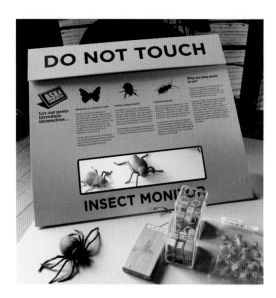

screen thus allowing the speaker to be able to explain to large groups or individuals.

Each part of the exhibit can be adapted to the needs of the different palaces in order to explain the unique challenges of combating pests at each site.

The portable exhibit has been a great success in engaging the public at Kensington Palace and Hampton Court and will be used in summer 2011 at the Banqueting House and the Tower of London. The exhibit is often on display at half-term and summer holidays. It has led to stronger links with the Education and Outreach department, and is often praised as an interesting and informative addition by younger and older visitors alike.

Historic Royal Palaces is the independent charity that looks after the Tower of London, Hampton Court Palace, the Banqueting House, Kensington Palace and Kew Palace.

A database devoted to the insects of cultural heritage

Fabien Fohrer*, Michel Martinez† and Franck Dorkeld†

**Centre Interrégional de Conservation et de Restauration du Patrimoine (C.I.C.R.P.),
Marseille, France*
*†Scientific partners, Institut National de la Recherche Agronomique (INRA),
Centre de Biologie pour la Gestion des Populations (CBGP), Montpellier, France*

The Interregional Centre of Conservation and Restoration of Cultural Heritage of Marseille (CICRP) and the National Institute of Agronomic Research (INRA), of Montpellier in collaboration with the National Centre of DNA sequence (GENOSCOPE) of Evry, developed a database devoted to the 'Insects of Cultural Heritage'.

Species of studied insects

The database inventories and describes the damaging insects present in heritage places constituting a risk for the conservation of the cultural property in museum collections, libraries and archives and historic buildings.

The objective of the creation of such a database

The database is a tool allowing survey and identification of the insects present in these places, this data is essential for preventive and curative measures to undertake in case of infestation.

The target users

It is dedicated to any professional in charge of cultural heritage conservation and to any other professional and scientist concerned or interested by these questions. Its contents, ergonomics and presentation make the database accessible to any user.

Components of the database

For each species several entries are proposed, such as:

- the common names
- the scientific names
- the infested materials
- the molecular sequence of DNA
- the keys of identification (dichotomic and visual).

These entries are related to a descriptive form for each species and give information about:

- their biology
- their frequency
- their distribution
- their danger to cultural heritage materials.

Examples of two species of insect present in the database are shown in Figures 1 and 2.

Reference

www.montpellier.inra.fr\CBGP\insectes-du-patrimoine

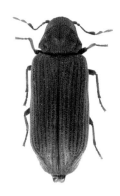

Fig 1
Anobium punctatum *(De Geer, 1774).*

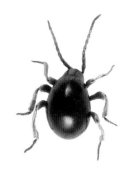

Fig 2
Gibbium psylloides *(Czempinski, 1778).*

The English Heritage integrated pest management programme

Dee Lauder

English Heritage, Conservation Team, 1 Waterhouse Square, Holborn, London, EC1N 2ST, UK,
dee.lauder@english-heritage.org.uk

English Heritage is responsible for a vast range of collections ranging from fine art to archaeology which have a direct link to over 400 sites. They are located in approximately 140 sites throughout England including large and small historic properties, purpose-built museums and stores.

Why use an IPM programme?

A well planned and executed integrated pest management (IPM) programme prevents problems occurring and prevents crisis in a collection. In order to develop an IPM strategy it is important to understand and recognise some of the key components of successful pest control with the participation of staff. These are:

- avoid pests by keeping pests out
- prevent pests by denying them safe haven
- recognise the main species of pest and the damage they cause
- assess the problems by inspection and trapping
- solve pest problems by improving the environment and carrying out appropriate treatments
- review IPM procedures changing them when necessary to improve the strategy.

Where do we implement?

To date we currently have an IPM programme in approximately 72 sites. These are sites where we house vulnerable collections, for example textiles, wood, paper, leather and silk that can be attacked by insect pests. The total number of insect monitoring traps can range from 77 traps at Brodsworth Hall, Yorkshire, to 7 traps in Bishops Waltham Palace, Hampshire.

What do we do?

We prevent damage to our collections in our houses and stores by setting up a personalised and manageable IPM system for each site. The main principles are:

- monitoring for insect pests
- targeting treatment only where it is needed
- modifying the environment to discourage insect attack.

Setting up a monitoring system

Place sticky museum traps and recording trap locations on site floor plans.

- location – against walls and in fireplaces
- number and date each trap
- use a realistic number of traps.

Monitoring and analysis

Check traps four times a year – spring, summer, autumn and winter.

- use a hand magnifier, torch, new traps, record sheet and rubbish bag
- two people make it easier

- examine traps in the rooms or gather them together
- smaller properties will take a morning, larger sites may take a day or two
- only record the insect pests
- record 'clear' if the trap is empty.

Plot results on a quarterly monitoring sheet and house plan over a year resulting in a yearly site report (Fig 1).

Benefits of an IPM programme

- indicates early warning of possible infestation so preventing damage
- encourages staff awareness and participation
- looks at the 'whole picture' rather than reacting to an individual crisis
- implements a well-planned and manageable programme at each site to prevent serious problems from developing

- targets control using safe methods to achieve better results at a lower cost rather than large-scale remedial treatments
- ensures that IPM is relevant to the needs of the building as well as the collections
- uses other sources, such as local information and expertise

Managing the IPM programme

The Collections Pest Control Manager oversees the training and development of skills and practices and provides ongoing support to staff which includes:

- pest identification training and basic treatment training
- positive communication between the Collections Conservation Team, site staff and Property Maintenance staff
- providing advice to external organisations and people.

Fig 1
House plan Brodsworth Hall.

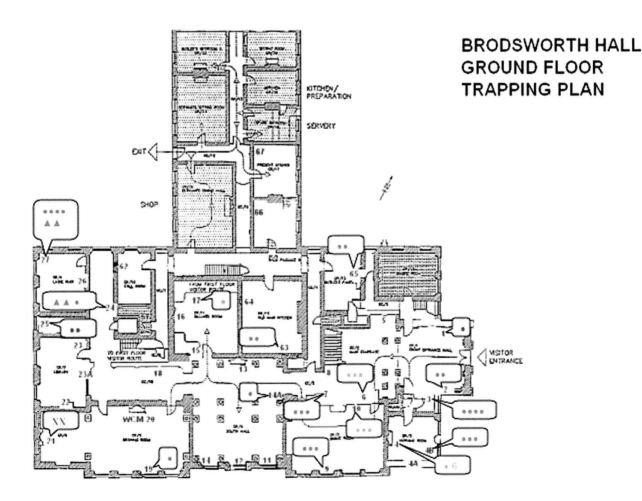

BRODSWORTH HALL GROUND FLOOR TRAPPING PLAN

The Exosex clothes moth system and English Heritage

Dee Lauder

English Heritage, Conservation Team, 1 Waterhouse Square, Holborn, London EC1N 2ST, UK,
dee.lauder@english-heritage.org.uk

English Heritage (EH) is responsible for a wide range of collections ranging from fine art to archaeology. They are responsible for approximately 140 sites throughout England including large and small historic properties, museums and purpose-built stores.

What is the Exosex clothes moth system?

Exosex is a non-chemical and non-toxic 'pest confusion' treatment designed specifically to control the highly destructive larvae of the webbing clothes moth. It uses a synthetic female pheromone to attract male clothes moths into a dispenser where the Exosex 'Entostat' powder combined with the pheromone is situated. Males are lured to the dispensers and the powder coats their bodies. As a result the senses of the coated moths are overwhelmed and they cannot detect females. They then attract other male clothes moths and so spread the confusion effect. This means that the females do not mate and they lay non-viable eggs resulting in far fewer larvae.

Where did we implement the system?

We implemented the system at Marble Hill House, London, after an increase in clothes moth numbers was recorded in 2004 and 2006.

This continuing increase caused concern so we took the decision to start a trial of the Exosex moth confusion technique in July 2007.

Setting up the Exosex system

We had to show discretion as to where the Exosex dispensers were placed as the house is open to the public at weekends and is also used for private functions. The new CL tab hanging dispensers (Fig 1) are much smaller than the

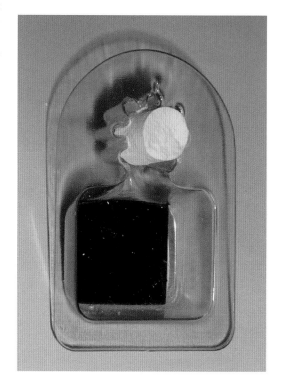

Fig 1
Exosex CL tab dispenser.

previous CLM designs and as a result we have found them to be easier to site under items of furniture, against walls, behind screens, on top of four-poster beds, in fireplaces and on stairwells. We currently use 24 dispensers throughout the house in various locations.

Monitoring and analysis

We agreed to keep the monitoring and recording of this system 'in-house' by our own staff as they are trained and experienced in insect pest identification, trapping techniques and basic treatment methods. They received initial training from the external pest control technicians so that they would be responsible thereafter, every six weeks, for replacing the Exosex CL dispensers. This allowed us to determine from our own quarterly monitoring trapping records whether the system is effective in decreasing and stabilising the number of clothes moths being caught on the traps. Results just over three years later have shown that the system has been effective. The moth counts have decreased with a total of 2 being caught during 2010 compared to 41 in 2004 and 36 in 2006. This result is very encouraging and we are continuing the trial in 2011 (Fig 2).

Benefits of the Exosex system and in-house deployment

- There has been a significant decrease in the amount of clothes moths being caught on traps without using pesticides or other treatments.
- There is now less threat to the collections from insect pests.
- EH staff undertake and continue the dispenser changes along with the tried and tested in-house IPM monitoring system.
- There is a saving in costs as we do not pay for external technician fees.
- If an external technician was used to replace the dispensers a member of the in-house staff, for security reasons, would have to accompany them. By doing this ourselves, we do not have this problem.
- We have control of the data for writing up a yearly report and publication.

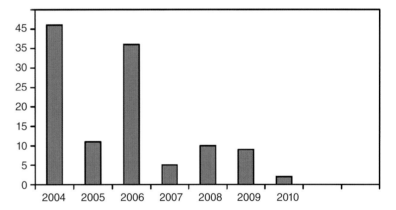

Fig 2
Webbing clothes moth catch on sticky traps

Using parasitoid wasps (*Lariophagus distinguendus* and *Trichogramma evanescens*) in integrated pest management against beetle and moth infestation: a preliminary evaluation

Pascal Querner*† and **Stephan Biebl‡**

University of Natural Resources and Applied Life Sciences, Department of Integrated Biology and Biodiversity Research, Institute of Zoology, Gregor-Mendel-Straße 33, A-1180 Vienna, Austria, pascal.querner@boku.ac.at (author for correspondence)

†Rathgen Research Laboratory, National Museums Berlin, Stiftung Preußischer Kulturbesitz, Schloßstraße 1 a, D-14059 Berlin, Germany

‡Pest Control Consultant, Mariabrunnweg 15, D-83671 Benediktbeuern, Germany

Biscuit beetles, *Stegobium paniceum*, and webbing clothes moths, *Tineola bisselliella*, cause a large amount of damage to museum objects all over the world. Some organic objects and materials are very attractive to these two species and collections are often re-infested after treatment. Both species are common in cities where they also occur in private homes. In the last few years, parasitoid wasps were applied in biological pest control strategies to reduce infestations of insect pests in mills and storage in the food industry (Schöller 2010; Schöller *et al* 1997). Parasitoid wasps are usually host-specific and lay their eggs on the larvae and eggs of their host species. We present observations from different museums in Germany and Austria where parasitoid wasps were used, firstly, as a biological pest management strategy and, secondly, as preventive management: *Lariophagus distinguendus* against *Stegobium paniceum* and *Trichogramma evanescens* against *Tineola bisselliella*.

To control active biscuit beetle infestations, 1,200 *Lariophagus distinguendus* wasps were released in a book library in Germany (Augsburger Stadtarchiv), 1,080 amongst paintings lined with starch paste linings in a store in Vienna (Kunsthistorisches Museum) and 9,000 in two ethnological collections in Berlin (Ethnological Museum) (Figs 1 and 2). The wasps reduced the population of the biscuit beetles (measured by monitoring over the following year), but this reduction was hard to differentiate from other pest control measurements such as by nitrogen treatment. 36,000 *Trichogramma evanescens* were released in the Technical Museum Vienna, Austria, and 45,000 in the Deutsches Museum Verkehrszentrum Munich, Germany, to control clothes moth. Active clothes moth infestations are harder to treat with these parasitoids because the wasps are smaller and only parasitise the eggs. But with a

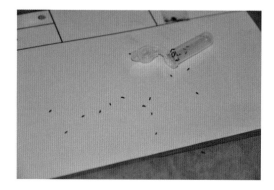

Fig 1
Release of Lariophagus distinguendus *parasitoid wasps in an ethnological collection in Berlin infested by biscuit beetles* (Stegobium paniceum). *(Photo by Pascal Querner)*

very regular exposure to wasps (three years plus), the clothes moth populations can be reduced over the years. Significantly fewer pests were found in the historic cars in the following years. Both species of parasitoid wasps can be applied as a preventive strategy in museum stores if climate conditions are right and resources are available. Using parasitoid wasps is still new in integrated pest management (IPM) in museums, and we suggest that it should be used alongside regular insect monitoring and other preventive measures. Research is still needed in this field of IPM to evaluate the efficiency of the wasps, the distance over which they can actively find their hosts and the number of individuals and times per year that the parasitoids should be released.

Conclusion

Parasitoid wasps have potential against insect pests in IPM as the wasps are cheap (a few hundred euros per treatment), easy to release in confined areas and can be applied as a preventive measure alongside an insect pest monitoring programme. More research is needed to critically evaluate their use against active infestations.

Acknowledgements

We thank the above-mentioned museums for allowing us to work with parasitoid wasps in their collections.

References

Schöller, M, Prozell, S, Al-Kirshi, A G, Reichmuth, C 1997 'Towards biological control as a major component of integrated pest management in stored product protection'. *Journal of Stored Products Research* **33**, 81–97

Schöller, M 2010 'Biological control of stored-product insects in commodities, food processing facilities and museums' in Carvalho, M O, Fields, P G, Adler, C S, Arthur, F H, Athanassiou, C G, Campbell, J F, Fleurat-Lessard, F, Flinn, P W, Hodges, R J, Isikber, A A, Navarro, S, Noyes, R T, Riudavets, J, Sinha, K K, Thorpe, G R, Timlick, B H, Trematerra, P, White, N D G (eds) *Julius-Kühn Archiv Proceedings 10th International Working Conference on Stored Product Protection*, 27 June to 2 July 2010, Estoril, Portugal, **425**, 596–606

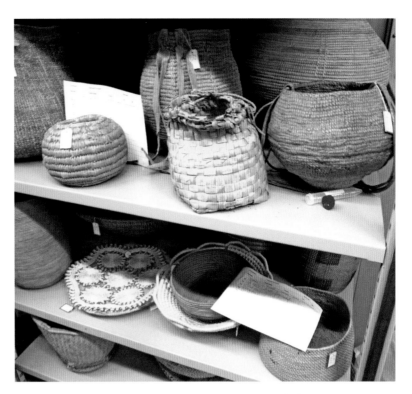

Fig 2
Lariophagus distinguendus *parasitoid wasps were released into this ethnological collection in Berlin which was infested by biscuit beetles* (Stegobium paniceum). *(Photo by Pascal Querner)*

Developing and implementing an integrated pest management concept in the Georgian National Museum in cooperation with the National Museums in Berlin

Pascal Querner*†, Nino Kalandadze‡,
Vakhtang Khoshtaria‡, Vera Phakadze‡, Manfred Nawroth‡
and Stephan Simon†

*University of Natural Resources and Applied Life Sciences, Department of Integrated Biology and Biodiversity Research, Institute of Zoology, Gregor-Mendel-Straße 33, A-1180 Vienna, Austria, pascal.querner@boku.ac.at (author for correspondence)
†Rathgen Research Laboratory, National Museums Berlin, Stiftung Preußischer Kulturbesitz, Schloßstraße 1 a, D-14059 Berlin, Germany
‡Georgian National Museum, Purtseladze Street, GE-0105 Tbilisi, Georgia

The Georgian National Museum is a complex of ten museums and two research institutes. Together they constitute a rich network devoted to the fine arts, archaeology, ethnography, and natural history. The Georgian National Museum and the National Museums of Berlin are part of a European Union Twinning Project (2010–12). The aim of the project is to support the institutional development of the Georgian National Museum in Tbilisi and other sites in the country.

The project started in June 2010 and at the beginning sticky blunder and pheromone traps for clothes moths were placed in the oriental collection of the Janushia Museum and the Fine Arts Museum (with a total of 230,000 objects) in Tbilisi. The Oriental Art Collection (5,000 objects) is exhibited and stored in 370m^2 under unfavourable conditions (temperature 8–29°C and 59–80 per cent relative humidity (RH)) and is a pilot case study in the project to evaluate the collection condition, implement an integrated pest management (IPM) programme, and relocate the collection to a new storage site. The valuable collection contains many different materials including a large carpet and textile collection, wood objects, an Egyptian mummy, paintings and drawings (Figs 1 and 2).

In August 2010 traps were checked for the first time and six pest species were found in the exhibition spaces of the Oriental Art Collection and storage areas: case-bearing clothes

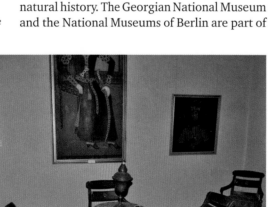

Fig 1
Exhibition room setting Oriental Art Collection, Fine Arts Museum in Tbilisi.

moths (*Tinea pellionella*) and webbing clothes moths (*Tineola bisselliella*), biscuit beetle (*Stegobium paniceum*), the common furniture beetle (*Anobium punctatum*) and carpet beetles (*Attagenus* and *Anthrenus*).

To treat the infested objects a nitrogen tent will be built in the museum and a freezing chamber will be bought. Other problems in the collection are the bad state of the building, climate condition, lack of regular cleaning and housekeeping and the fungal growth on objects such as carpets or mummies.

Both museums in Berlin and Tbilisi work together to implement an IPM concept in the collections and to train museum staff. In the future one person in the Georgian museum will be responsible for monitoring and a second person will treat objects with mobile nitrogen tents. In 2011 other collections will be monitored with insect traps to get a better insight into the presence of pest species.

Conclusion

We have shown that even during a short period of pest monitoring a large number of species can be detected and problems evaluated. More collections of the Georgian National Museums will be sampled in the future and a workshop on IPM will be held during 2011.

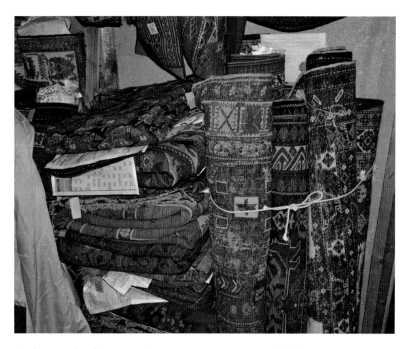

Acknowledgement

We thank the EU for funding this Twinning project.

Reference

http://museum.ge/web_page/index. php?id=192 (accessed 16 September 2011)

Fig 2
Storage area in the Oriental Art Collection, Fine Arts Museum in Tbilisi.

Saving money with integrated pest management at the Kunsthistorisches Museum, Vienna

Pascal Querner*†, Michaela Morelli‡, Elke Oberthaler‡, Monika Strolz‡, Katja Schmitz von Ledebur‡, Isabelle Zatschek‡, Regina Hölzl‡ and Irene Engelhardt‡

*University of Natural Resources and Applied Life Sciences, Department of Integrated Biology and Biodiversity Research, Institute of Zoology, Gregor-Mendel-Straße 33, A-1180 Vienna, Austria, pascal.querner@boku.ac.at (author for correspondence)
†Rathgen Research Laboratory, National Museums Berlin, Stiftung Preußischer Kulturbesitz, Schloßstraße 1 a, D-14059 Berlin, Germany
‡Kunsthistorisches Museum Vienna, Maria Theresien-Platz, A-1010 Vienna, Austria

The Kunsthistorisches Museum in Vienna is one of the largest fine arts collections worldwide. It comprises the Kunsthistorisches Museum with its different collections, the Austrian Theatre Museum and the Museum of Ethnology. The Kunsthistorisches Museum was the first

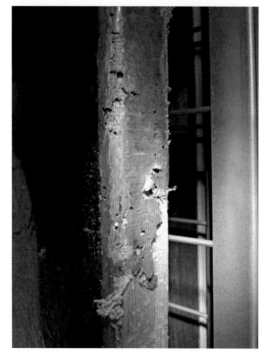

Fig 1
Biscuit beetle (Stegobium paniceum) infesting tacking edge of a painting, with starch paste linings at the Kunsthistorisches Museum, Vienna.

museum with integrated pest management (IPM) in Vienna and is still the only museum with a nitrogen chamber. For the relocation of the main storage depot of the museum not all the collections could be treated as a preventive measure. Therefore during the planning and building of the new storage depot (2009–11) all relevant collections were monitored as part of a large IPM programme for the museum.

Through this monitoring and from the results of previous years in other collections, some infested objects were found. Specific pest problems in the collections of the Kunsthistorisches Museum are biscuit beetle (*Stegobium paniceum*) infestation of paintings lined with starch paste linings (Fig 1). Webbing clothes moth (*Tineola bisselliella*) were found in the Museum of Carriages and in a storage site of the Austrian Theatre Museum (Fig 2).

In order to prevent an introduction of insect pest species into the new storage depot, infested collections will be treated with nitrogen before the move of the collections in the summer of 2011. A large nitrogen tent will be built for the treatment of the paintings on the old storage site. Smaller tents will be built in the storage depot of the infested theatre collection, and for the horse carriages. In the new storage depot a new monitoring programme

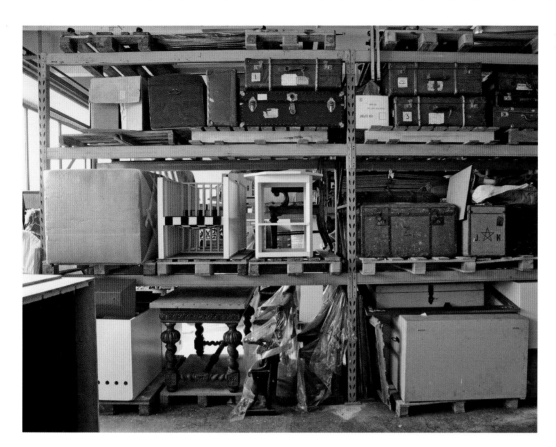

Fig 2
Storage site of the Austrian
Theatre Museum, Vienna.

will start in 2012 to monitor the success of the treatments. We can show that with a good IPM and monitoring programme not all collections have to be treated before relocation, which considerably reduces the costs for pest treatments. Pest problems can be dealt with during the construction of the new storage site.

have to be treated before removal to a new storage site. The longer monitoring of insect pests is in place, the better and more reliable is the result. We therefore recommend that all large and small collections and museums introduce a monitoring and IPM programme.

Conclusion

An integrated pest management programme can help to save money as not all collections

Acknowledgement

We thank the Kunsthistorisches Museum in Vienna for supporting and funding an IPM programme in the different collections.

The importance of support and involvement of the whole institute for successful IPM

Suzanne Ryder

The Natural History Museum, Cromwell Road, London, SW7 5BD, UK, s.ryder@nhm.ac.uk
www.nhm.ac.uk

Collections are continually under threat of irreversible damage by insect and rodent pests. The Natural History Museum (NHM) introduced a comprehensive integrated pest management (IPM) programme in 2002. This was in response to changing health and safety legislation, banning many of our methods of pest control, for example dichlorvos (2,2-dichlorovinyl dimethyl phosphate, DDVP), and to be in accordance with the NHM environmental policy to seek an alternative to pesticides.

The principles behind IPM are to eliminate, or at least limit, any possible chance of entry or contamination by a process of:

- quarantine
- monitoring
- housekeeping
- facilities design and maintenance
- training
- storage
- environmental control
- best work practice.

The aim of the IPM strategy is to protect the collections against pest attack using all of these essential elements and this can only be achieved by a holistic approach. This needs to involve everyone in the Museum contributing to its success with key involvement from conservation, management, the Science group, collections staff and Estates management and housekeeping (Fig 1).

After almost 10 years the NHM can report that our approach to IPM is a success. It is clearly something that cannot be written off as a job well done but the consistent message is having a positive impact. This success lies with getting everyone on board with effective training, especially key departments like housekeeping and Estates. The efforts and involvement of these departments have proved invaluable and it is essential that it be maintained.

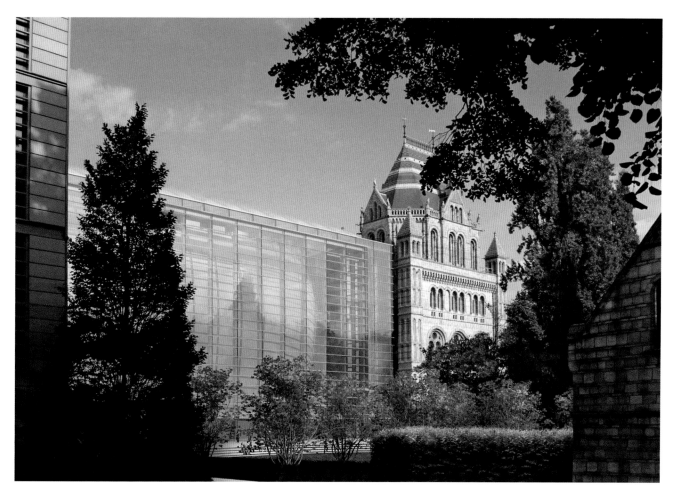

Fig 1
The purpose-built
collections facility; the
Darwin Centre joins the old
Waterhouse building at the
Natural History Museum.

Defining success: The Colonial Williamsburg Foundation's new institution-wide IPM programme

Patricia Silence* and Ryan Jones†

Colonial Williamsburg Foundation, P.O. Box 1776, Williamsburg, Virginia, 23187-177, USA,
www.history.org/foundation/index.cfm
*Conservator of museum exhibitions and historic interiors, psilence@cwf.org
†Integrated pest management technician, rjones@cwf.org

Fig 1
An aerial view of the
Historic Area at Colonial
Williamsburg.

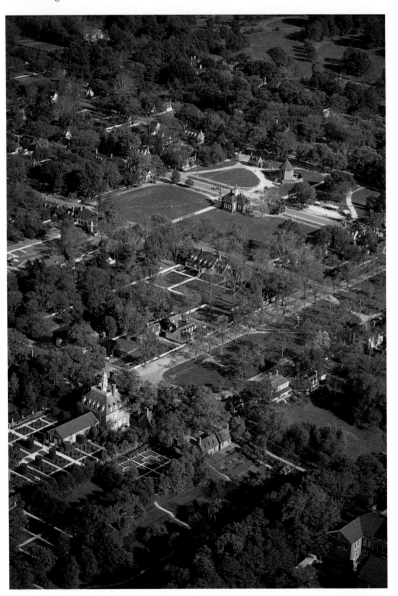

Background

The Colonial Williamsburg Foundation (CWF) is the oldest and largest outdoor living history museum in the United States. Since its inception, Colonial Williamsburg has preserved and presented American history and heritage. The site includes over 800 buildings and structures situated on more than 300 acres (120ha). Collections include archaeological material, decorative arts and other cultural heritage items (Fig 1).

For many years, CWF required the use of professional pest services to assist with the preservation of collections, structures, landscapes, and rare breeds as well as ensuring visitor safety. Early in 2009 a group was formed to outline pricing requests to a small group of pre-selected integrated pest management (IPM) contractors and to determine a course of action. This committee was the foundation of our current in-house IPM group, and included representatives from facilities maintenance, landscape (including a certified pest applicator), collections, and finance.

As we evaluated what a successful programme would entail, it became clear that our definition of success should be the same as our IPM provider's – we wanted to prevent damage to collections and structures, use as little pesticide application as possible, and foster a sense of ownership in the programme among employees working for the Foundation. Due to the size and scope of the work necessary, we determined that an in-house approach would be the most sensible course of action in developing an effective IPM programme.

Since its implementation in 2010, our IPM programme has been successful. In addition to carrying out detailed on-site inspections and identifying current or potential pest issues, Colonial Williamsburg currently uses its network of nearly 3,500 employees to broaden the scope and reach of the programme.

Departmental cooperation

- Detailed pest inspection reports generate work-orders that are completed by maintenance and support teams. Structural repairs are carried out to eliminate conditions conducive to pest activity, thus reducing repetitive chemical applications.

- Good housekeeping practices are encouraged in areas where food or beverages are consumed. Site management teams are provided with preventive recommendations for improving sanitation and storage procedures.

In-house IPM training

- CWF provides annual state-approved IPM recertification training for over 40 in-house applicators. This improves knowledge concerning seasonal pest pressures and increases awareness of what is, or is not, a genuine pest issue.

- Conservation technicians who care for historic buildings are trained in the basics of IPM inspections and practices. This enables them to recognise the early signs of pest activity and understand the rationale of trap and monitor placement.

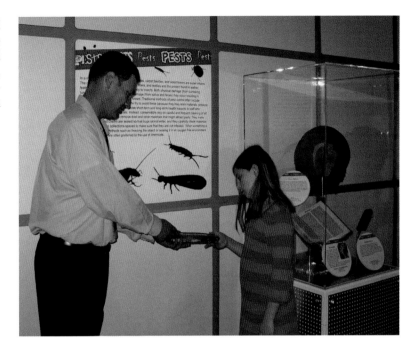

Communication about the IPM programme for employees and guests

- Web-based information is maintained about industry trends, products, and pests. This is shared with staff members. Inquiries from employees are researched and answered in a timely manner.

- A tour entitled 'When insects invade: the unwanted museum visitor' is offered as part of a current conservation exhibit. It highlights the basics of museum IPM. Originally designed for children, this activity in the DeWitt Wallace Museum is attended and enjoyed by visitors of all ages (Fig 2).

Fig 2
Ryan Jones discussing wood destroying pests with a young visitor.

Strategies against insects in the galleries and stores of the North America Native Museum, Zurich

Jenny Studer* and Artemi Chaviara†

*MA Preventive Conservator, Zurich, Switzerland, jenny.studer@network.rca.ac.uk
†MA Preventive and Object Conservator, Athens, Greece, chaviaraa@gmail.com

Preventive conservation strategies on integrated pest management (IPM) have recently been introduced in the North America Native Museum (NONAM) in Zurich, Switzerland. Several procedures are being applied to protect closed and open displayed objects in the museum galleries and storerooms.

A large number of the exhibits of the permanent collection are on open display, some intended to be touched by visitors interacting with the collection. The variety of the organic materials, such as wood, leather, feather, fur, straw, textiles and taxidermy, on open display makes the objects vulnerable to insect infestation.

In April 2010 the NONAM opened two temporary exhibitions one on glass bead trading between Europeans and Native American Indians and the other on contemporary art from Native American artists. Pine needles from a forest, forming a thick layer on the floor indicating a carpet, was part of Hannah Claus' installation 'Pine' in the contemporary art exhibition. Visitors being able to walk on the carpet was the vision of the artist (Fig 1).

The pine needles were exhibited without being checked or treated. In combination with the open access of the public to the installation, it was realised that this medium presented a risk and could be a means of spreading insect populations which would be a threat to objects in the collections. This led the museum to consult a preventive conservator on the IPM strategies for the permanent collection of the NONAM. This led to spraying of the pine needles with an insecticide as a preventive measure.

Checking of the galleries, the displays and some of the objects for a first assessment for insect evidence in the museum was the primary action. The general monitoring of the museum and storage rooms was decided after finding a few adults, larvae and casts of *Attagenus smirnovi* in two displays with vunerable materials. An insect monitoring programme was started by placing traps in the galleries, some displays and the stores of the NONAM for more accurate assessment and confirmed the presence of pest activity. Most larvae and adults were found in a soundproof space, a completely dark room covered with foam plates and grids on the floor, where visitors listen to the sounds of wild animals of North

Fig 1
Hannah Claus, 'Pine'
installation, 2010.

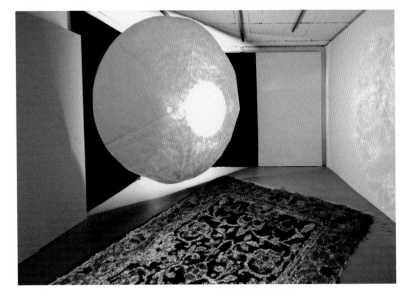

America. The gaps of the grids are conducive to the accumulation of dust, dirt, clothing fibres, hair and skin particles from visitors. This debris provides an ideal environment for insect activity. The room was dismantled, the surfaces cleaned and sprayed with Constrain and the edges of the floor were treated with Bioshield, a natural, non-toxic product. Further monitoring showed that these measures have greatly reduced the insect activity (Fig 2).

The project is still in progress and future actions, such as a survey of the vulnerable objects in the museum, an inspection and a surface cleaning of certain objects, will take place when more results of insect activity are gathered from the current monitoring.

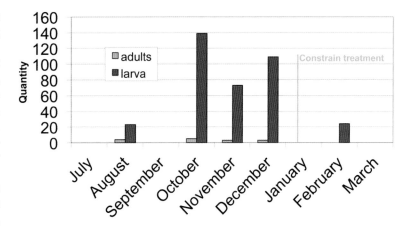

Fig 2.
Vodka beetle findings from August 2010 to February 2011.

Declaring war on clothes moths

David Thomson

Senior Preventive Conservator, Glasgow Museums, Glasgow, Scotland, UK,
david.thomson@glasgowlife.org.uk

Introduction

The common or webbing clothes moth (*Tineola bisselliella*) is a tough opponent. In the UK today, the species is exploiting opportunities to live indoors in households and public buildings, including museums. There are many factors in the moth's favour. For example: some museum display case designs do not allow effective floor-level cleaning, cleaning contracts do not always tackle areas where moths can breed, and some insecticides have been banned. Also, reports of domestic infestations have increased, which may be related to climate change.

Questions

- Will we ever completely eradicate moths from museum buildings?
- How can we minimise damage and make our buildings and collections less attractive to clothes moths?
- Do we have the right tools to do the job?
- How can the design of future museum buildings and display systems contribute to pest control?

Discussion

It seems likely that current infestations will not be eradicated in the short term. The use of traps and insecticides will help control levels. Better deterrents and insecticide treatments would help. Cleaning and housekeeping is vital to control infestations. Many museum displays, including display cases, furniture and fittings, plinths, seating and interactives make housekeeping either impossible or very time-consuming. The future design of museum furniture and fittings is a very important factor.

Is your museum cleanable?

Does your current method of cleaning sweep dust under display cases, into seams between floorboards, under skirting boards, into underfloor voids through gaps in floorboards? Moths can live on dust because it contains protein, for example wool from clothes, and skin particles (Fig 1). Does your current cleaning regularly reach these areas? This is particularly important if woollen carpeting is in use (Fig 2). The need for good cleaning applies to all areas in the museum, including offices, stores, displays, cafes, attic spaces, service areas, dead areas under stairwells and so forth.

New display cases?

If you are designing or commissioning new display furniture, is it cleanable? Or are you introducing a potential dust trap to the museum?

Anything introduced to a museum that prevents floor-level cleaning can potentially create a moth breeding ground. For example, a large display case on a plinth style base, with a 'shadow gap' at floor level, will create a dust trap. Such a space is unlikely to be disturbed, and there will be very little air movement. This environment will support all stages of the clothes moth life cycle: eggs, larvae, pupa, and allow adults to hatch and move on to spread the infestation. Dropped food and general detritus have been found to end up under such cases, far out of the reach of cleaning equipment.

Action taken

The problems identified have required a variety of responses: objects have had freezer treatment, parts of buildings have been chemically treated, staff have been trained to be vigilant and apply simple treatments. Cleaning methods are under review. Extensive experience in the use of pheromone insect traps has been gained. The design brief for future museum projects takes account of the need for cleaning.

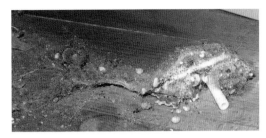

Fig 1
Larvae activity in floor-level dust and detritus.

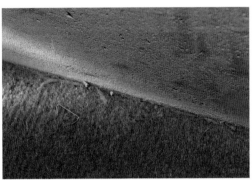

Fig 2
Clothes moth webbing on woollen carpet.

'Have I got a pest problem?' Mapping insect vulnerability of costume collections

Efstratia Verveniotou* and Jane Henderson†

*Natural History Museum, London, England, UK, e.verveniotou@nhm.ac.uk
†Cardiff University, Cardiff, hendersonlj@cardiff.ac.uk

Introduction

This poster presents a simple software tool for mapping the risk of pest infestation in costume collections. Users are prompted to consider factors that contribute to the susceptibility of a costume collection to insect infestation. Those responsible for costume collections that have not yet employed integrated pest management (IPM) can use the programme to quickly and easily identify areas of high risk and undertake actions to reduce them.

Methodology

The software has been developed to bring together and evaluate information from current literature on how the nature of a textile, insect biology, museum activities and display or storage conditions can affect insect activities. The use of past surveys, pilot studies, case studies, published research on the eating habits of various insects and current practices in the care of costume collections help profile the high-risk areas.

Factors examined

The insects

To inform the evaluation of priorities the project examined the insects most likely to attack museum textiles. Insect biology helps provide information on insect life cycle (eating, living and breeding habits), their favoured environments (temperature, relative humidity, light and pollution), along with how the insects can be monitored (sighting, damage caused, and life-cycle residues).

The collection

The extent to which textile-eating insects thrive depends on the nature of the textiles. Factors related to the composition, dyeing, finishing and sizing of textiles, along with decoration and design of the costumes, can attract or help insects flourish. Soiling and surface dirt from the accumulation of dust, oils or food spills are also evaluated.

The care of the costume collections

The way collections are used and how costume collections are cared for provide information on factors that may promote, sustain and allow insect infestations to remain undetected. The building fabric along with materials and designs of display and storage methods for costumes are examined. Collection care practices concerning loans, routine housekeeping, monitoring and seasonal cleaning are also considered.

Software application

Using all of the features described, the authors formulated an interactive software application to help museum staff map the areas of their collection that are vulnerable to insect infestation. The software collects information through

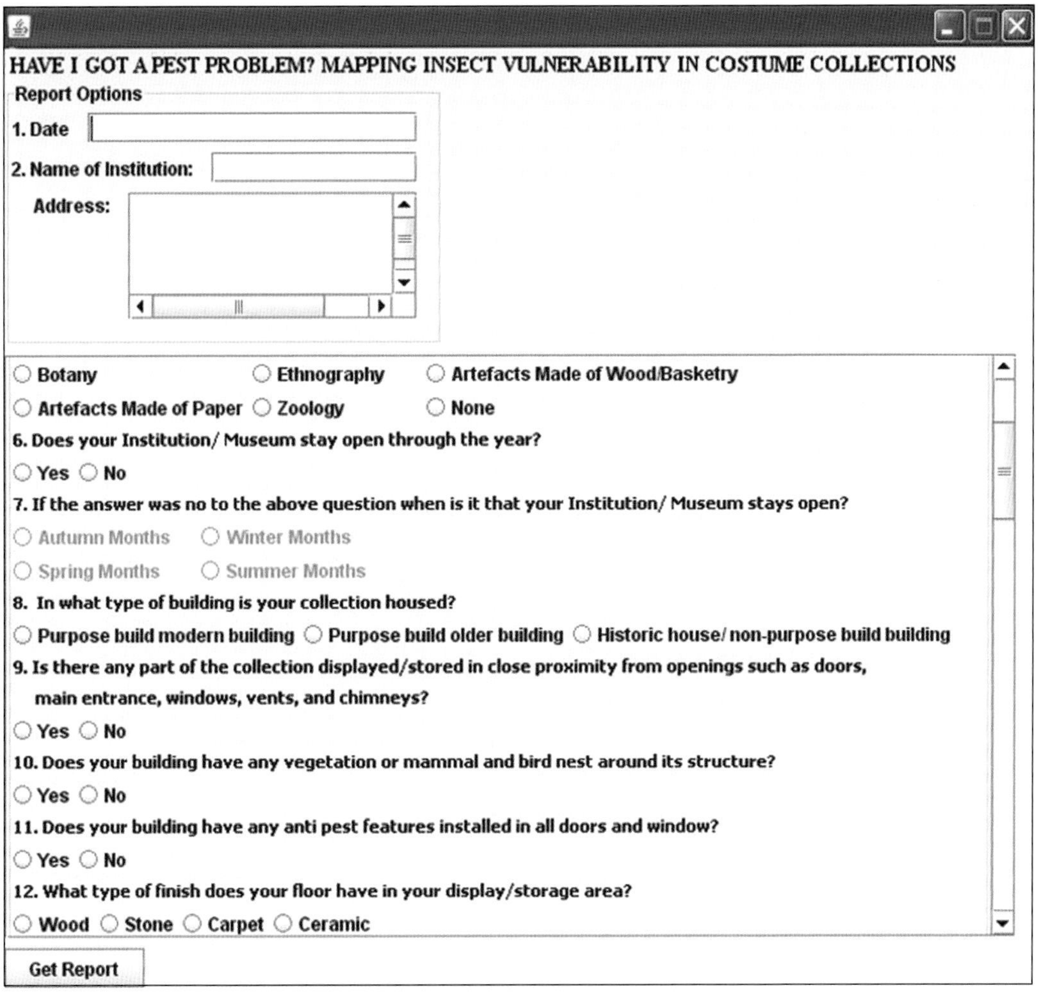

Fig 1
Screenshot of running application, 'The Questionnaire'.

a set of multiple choice questions and builds them into a report on how current museum practices can affect insect activity (Fig 1). Additionally, it provides the user with a set of immediate actions that should be considered and information on how to obtain further help on developing an IPM Strategy.

Acknowledgement

The authors would like to thank Ros Witford from the Imperial War Museum for her helpful comments and discussion at the first stage of designing this project.

APPENDIX: MATERIALS AND SUPPLIERS LIST

The names and details of suppliers of products and materials included in the list below were provided by the authors of the papers and posters presented at the 2011: A Pest Odyssey, 10 Years Later conference, The British Museum, London, 26–28 October 2011.

Data loggers

Hanwell Instruments Limited
www.hanwell.com/ (accessed 16 September 2011)
(Hanwell Radiolog)

Hastings Data Loggers
www.hdl.com.au/ (accessed 16 September 2011)
(Tinytag View 2 temperature and relative humidity logger)

INTAB Interface-Teknik AB
www.intab.se (accessed 16 September 2011)
(Tinytag Ultra, temp/RH data loggers)

Insect traps and lures

Exosect Limited
www.exosect.com/ (accessed 16 September 2011)
(Exosex system, Exosex™ CLTab)

Historyonics
www.historyonics.com/ (accessed 16 September 2011)
(Sticky (blunder) traps, pheromone traps, Constrain insecticide, Agrodust desiccant dust)

Killgerm Chemicals Ltd
www.killgerm.com/ (accessed 16 September 2011)
(Demi-Diamond AF Trap, webbing clothes moth and Indian meal moth pheromone traps)

PestimoServices
www.pestimoservices.com/ (accessed 16 September 2011)
(Pheromone (Finicon) traps)

Pre-Mal, Naturhistoriska riksmuseet
www.nrm.se/sv/ommuseet/samverkansparter/pre-mal.1045.html (accessed 16 September 2011)
(Sticky traps)

Insecticide

Anticimex AB Pest Control Company
www.anticimex.se (accessed 16 September 2011)
(Demand CS pesticide – a pyrethroid insecticide, active ingredient Lambda-Cyhalothrin. Permission to be used until 2010-12-31)

Biozida
www.biozida.ch/ (accessed 16 September 2011)
(Bioshield)

Historyonics
www.historyonics.com/ (accessed 16 September 2011)
(Sticky (blunder) traps, pheromone traps, Constrain insecticide, Agrodust desiccant dust)

Killgerm Chemicals Ltd
www.killgerm.com/ (accessed 16 September 2011)
(Demi-Diamond AF Trap, webbing clothes moth and Indian meal moth pheromone traps)

Microscope/endoscope

Maplin Electronics Ltd
www.maplin.co.uk (accessed 16 September 2011)
(Flexible snake scope (endoscope) camera)

Veho Europe
www.veho-uk.com (accessed 16 September 2011)
(Veho VMS-001, 20–200× magnification USB digital microscopes)

Pest control systems

Mobigas
www.mobigas.de/ (in German) (accessed 16 September 2011)
(Anoxic pest fumigation system)

Thermo Lignum® UK Limited
www.thermolignum.com/ (accessed 16 September 2011)
(Heat treatment for pest control)

Plastic film, sheeting, adhesive and gap-filler

AB Helmer Nilsson,
www.helmer-nilsson.se (accessed 16 September 2011)
(Transparent plastic tubing 300 × 0.09mm × 300m code: article 711359)

Hadleigh Enterprises Ltd.
www.hadleigh.uk.com/ (accessed 16 September 2011)
(Very high temperature (VHT) double-sided masking tape)

Livingstone International Pty Ltd
www.livingstone.com.au/ (accessed 16 September 2011)
(Blanket NBC Surviv-a-wrap Emergency blanket)

Schneidler Grafiska AB
www.schneidler.se (accessed 16 September 2011)
(Tyvek® 1443R, 1524mm × 100m, polyetelen)

STOPGAP
www.stopg-p.co.uk (accessed 16 September 2011)
(gap-filler for floor boards and skirtings)

Sundbybergs Textilcentrum
www.textilcentrum.se (accessed 16 September 2011)
(Insect netting, 'Voile plombé' width 300cm, with a band containing a chain of small lead weights sewn into a pipe-like fold, 100 per cent polyester, produced in Turkey)

Viper Packaging Pty Ltd
Norwest Business Centre, Baulkham Hills NSW 2153, Australia
No URL found
(Co-Ex Barrier Film 1300mm × 150μm)

Wurth UK Ltd
www.wurth.co.uk/ (accessed 16 September 2011)
(PUR adhesive)

Software

Adobe®
www.adobe.com/uk (accessed 16 September 2011)
(Adobe Photoshop and Acrobat software)

KE Software
www.kesoftware.com (accessed 16 September 2011)
(EMu collections management software)

Microsoft
office.microsoft.com/en-us/excel/ (accessed 16 September 2011)
(MS Excel)

OpenLayers
openlayers.org (accessed 16 September 2011)
(Software)

Training aids

English Heritage
www.english-heritage.org.uk/publications/historic-house-collections-management-plan/drawing-up-collections-management-plan.pdf (accessed 16 September 2011)
(Free poster A Helpful Guide to Insect Pests found in Historic Houses and Museums)

Northern States Conservation Centre (www.collection-care.org)
through www.museumclasses.org (accessed 16 September 2011)
(online IPM training)

Vacuum cleaner and HEPA filters

Eastern Shires Purchasing Organisation (ESPO)
www.espo.org (accessed 16 September 2011)
(Nilfisk UZ 964 hip-mounted vacuum and heap filters)